D1067498

The Versatile Satellite

The Versatile Satellite

Richard W. Porter

OXFORD UNIVERSITY PRESS
1977

Oxford University Press, Walton Street, Oxford OX2 6DP

OXFORD LONDON GLASGOW NEW YORK
TORONTO MELBOURNE WELLINGTON CAPE TOWN
IBADAN NAIROBI DAR ES SALAAM LUSAKA ADDIS ABABA
KUALA LUMPUR SINGAPORE JAKARTA HONG KONG TOKYO
DELHI BOMBAY CALCUTTA MADRAS KARACHI

ISBN 0 19 885104 9

Set by Hope Services, Wantage, Oxon.
and printed in Great Britain by
William Clowes Ltd., Beccles.

Preface

In the two decades since I began to work on the first satellite program of the United States, many hundreds of artificial earth satellites have been launched. Some have made possible important scientific observations, others have been used for applications of direct economic and social benefit, and yet others have served as barques to carry brave men out into space—from whence a few eventually travelled on to the moon and safely returned. The use of satellites has by now become so involved in the lives of so many different kinds of people in so many different ways that it seems important for almost everyone to have an understanding of what a satellite is, how it is put into orbit (and why it does not promptly fall down again), what special considerations enter into its design, and especially what it is 'good for' and why. I hope that this book will provide such an understanding.

Except for the Appendix, which is intended only for the reader who has studied a little elementary physics and calculus, the presentation is descriptive rather than mathematical, and is liberally illustrated with diagrams and pictures. Wherever it has seemed necessary to use technical jargon, the special terms are explained. The opening chapter contains a general historical background covering the early development of satellites from the viewpoint of someone who was there. In Chapters 4, 5, and 6, the continuing stories of communications, meterology, and navigation are reviewed in just enough detail to show how and why satellites have been so much used in these areas.

Not covered in this book are explorations of the moon and planets and interplanetary space missions. These are indeed fascinating subjects, but they could not be included here without making the book impossibly long or giving too short shrift to the versatile satellite, which in its own way is even more fascinating. Perhaps in another volume . . .

Richard W. Porter

Acknowledgements

FIGURES AND PLATES

4.1, 4.4, 5.3, 5.4, 5.10, 5.11, 7.6–7.9, 7.13–7.17, 10.1, 10.2, 10.4, 10.5, 11.1, 11.2, 11.3 NASA; 4.2 TRW; 4.3 USAF; 5.1, 5.18, 7.1, 7.19, 9.2 General Electric Space Division; 5.2, 5.19 RCA; 5.6, 5.7, 5.8, 5.12, 5.13, 5.14, 7.18 NASA, Courtesy General Electric Space Division; 5.9 Santa Barbara Research Center; 5.20 Philco Ford; 5.21 SOVFOTO; 7.10–7.12 Department of Interior, US Geological Survey; 9.3 Hale Observatories; 10.3 National Centre for Atmospheric Research, High Altitude Observatory, Boulder, Colorado.

Contents

1 Launching the first satellites

Spin along in spatial night
Artificial satellite.
Monitor, with blip and beep,
The Universe—and baby's sleep.
Frederick Windsor (1958) *The space child's mother goose*

There is something wonderful about looking down from a high place. Prehistoric men climbed tall trees and mountains. Medieval men built towers as tall as their engineering skills would permit; and it is not uncommon today for the President or Chairman of a large corporation to have his office on the highest floor of the tallest skyscraper his stockholders can afford. Envious of the birds, men learned to fly through the air higher and faster, although perhaps less gracefully, than the birds themselves, and in so doing achieved profound emotional satisfaction (ask any pilot) as well as great practical advantages. Baffled in their continuing quest for yet greater heights and speeds by the limitations of the atmosphere, men have now turned to an old Chinese invention—the rocket. Aimed straight up, even a relatively inefficient rocket can go higher than any aircraft—well beyond the 'sensible' atmosphere—and faster than any speeds that can be achieved by ordinary jet engines.

Suddenly or so it seemed, by the middle of the twentieth century, the time had come to try out an idea that had been thought about for many years: a vehicle that would fly high enough and fast enough to sustain itself indefinitely outside the earth's atmosphere by momentum alone, with no additional expenditure of energy; in other words, an artificial earth satellite. Just how one knew that the time had come is hard to say. Writers like Edward Everett Hale, Jules Verne, and Kurt Lasswitz had written about the subject in an imaginative way during the last decades of the nineteenth century. Stimulating though they were, however, these science-fiction tales were not to be taken seriously by scientists or engineers. Conceptual, and to some extent theoretical, studies during the first 35 years of the twentieth century by Tsiolkovsky, Oberth, Goddard, and others had convincingly shown that rocket propulsion could be developed, at least in principle, to the point where it could be used to launch an artificial satellite, but the technology available at that time was still woefully inadequate, and there seemed to be no very impelling reason for governments, corporations, or individuals to invest in rocket technology the kind of money that would be required.

German rockets

The big change came during the Second World War when the German Government, facing serious shortages of petroleum, aircraft construction materials, and pilots, decided to gamble on the promise of a small group of rocket experimenters, backed by an Army General, that they could produce a weapon that would deliver a ton of explosive warhead accurately at a distance of up to 300 kilometres, would travel at supersonic speed and so be very difficult to intercept, would require very little in the way of critical construction materials, would use industrial alcohol for fuel instead of petroleum, and would not require a pilot. Whether or not this was, in hindsight, a sound decision on the part of the wartime German Government is still debated by military experts. However, it was a decision that changed the world. The German engineers and technicians, working under General Walther Dornberger and Dr. Wernher von Braun, took a giant step in the technology of liquid-fuel rocket engines and automatic control of supersonic vehicles—a step which soon led to similar steps in the United States and the Soviet Union, revolutionizing ideas about strategic warfare and leading directly to the practical possibility of sending instruments and men into orbit around the earth and into interplanetary space.

Early American efforts and the IGY

By the early 1950s studies had been made under military auspices by American aircraft companies and by the prestigious Rand Corporation; all these studies indicated rather conclusively that a multiple-stage rocket launcher capable of putting a small satellite in orbit could be designed using only moderate extensions of existing technology. Meanwhile, less detailed studies by other groups, such as the British Interplanetary Society, were leading to much the same result. The American military studies were referred by the Secretary of Defense to the Guided Missiles Committee of the Research and Development Board. This Committee agreed that the technological feasibility of an artificial earth satellite had been adequately demonstrated, but added that if a project to build one were to be undertaken immediately it would be very expensive, that no application having a value commensurate with this cost had yet been determined (although many were somewhat vaguely anticipated), that the cost of developing a satellite launcher would undoubtedly drop during the next few years because of the extensive work being done to develop large military intercontinental ballistic missiles, and that it might be better to spend a few years actively studying how to use a satellite before establishing a project to build one. This advice was, in fact, accepted and implemented by the Department of Defense, which contracted with the Rand Corporation for a study of satellite applications.

During the years 1952–4 the American Rocket Society (which later merged with the Institute for the Aeronautical Sciences to become the American Institute for Aeronautics and Astronautics) carried out its own studies of the ways in which earth satellites could contribute to astronomy, biology, communications, geodesy, meteorology, and ionospheric physics. It also identified the National Science Foundation as the only governmental body in the United States then able to sponsor a scientific satellite project without any change in its existing charter. The Rocket Society issued a report proposing that the Science Foundation should undertake a satellite development project using public funds and looking to the Department of Defense for the launching capability. It seems likely that similar studies were being carried out at the same time in the Soviet Union, because in late 1953, A. N. Nesmeyanov, of the USSR Academy of Sciences, stated that satellite launchings were already feasible.

During the summer of 1954, an artificial earth satellite programme was recommended as part of the International Geophysical Year (IGY), which was to extend from July 1957 through December 1958. In October, 1954, the Comité Spécial de l'Année Geophysique Internationale (CSAGI) made a formal recommendation to its constituent national committees 'that thought be given to the launching of small satellite vehicles, to their scientific instrumentation and to the new problems associated with satellite experiments, such as power supply, telemetering and orientation of the vehicle'. This was the trigger that initiated action rather than words in the United States, and probably also in the Soviet Union.

The Department of Defense established a small Advisory Group on Special Capabilities (known as the Stewart Committee) to obtain a quick answer to the questions: Can we really launch a satellite in time for the IGY? Which of the alternative proposals for doing so should be adopted and why? It also required that the satellite launching must not interfere with any military rockets programmes or delay them, even to the extent of one day! By mid July 1955, the Stewart Committee had answered the first question affirmatively, and on this basis Dwight Eisenhower, then President of the United States, issued a statement to the effect that the United States intended to launch an IGY satellite. Then things really began to happen. The National Science Foundation, which was providing financial support for the IGY effort in the United States, obtained increased funds to build and instrument the satellites. A special panel of the United States National Committee for the IGY was established in the United States Academy of Sciences to manage the programme for the Foundation. The Naval Research Laboratory, whose proposal to launch the satellites using a modified *Viking* research rocket as the first stage had been selected as 'most likely to succeed' by a majority of the Stewart Committee, began work on a launcher of that configuration. It was also given responsibility for building and testing the satellite vehicles and for development, construction, and operation of a new radio interferometer tracking system. The Smithsonian Astrophysical Observatory was brought in to supervise the creation of a new telescopic camera of advanced design for photographing satellites against a star background and to organize and operate a worldwide network of these cameras. The Smithsonian Observatory also organized a world network of volunteer observers using simple optical devices and clocks, under the name *Moonwatch.* Scientists at universities or associated with industrial and government laboratories began developing miniaturized satellite experiments, from which the panel in the Academy would later select payloads for specific satellites. Dr. von Braun (who now worked in the United States at the Army's *Redstone* Arsenal), confident of his group's capability, continued to work on a launcher configuration based on the large but obsolescent *Redstone* military rocket, even without any official authority to do so.

Sputnik get there first

Meanwhile, in the Soviet Union, an Interdepartmental Commission on Inter-Planetary Communications had been established under the Academy of Sciences, and in April 1955 a local Moscow newspaper told of Soviet plans to launch a satellite. Unofficial statements were also made by knowledgeable Soviet scientists and engineers at international meetings during the autumn of 1955 and early in 1956. Finally, in September 1956, the USSR National Committee for the IGY officially announced its intention to launch a satellite during the IGY. Events soon made it clear that by this time Soviet scientists and engineers must already have been hard at work for many months.

As is not uncommon in such situations, there were problems, delays, and cost overruns in the approved American launcher programme. In 1956, the Stewart Committee, which had been called back to review the situation, recommended that the Army's programme be officially recognized and funded as a parallel or back-up effort. This recommendation was turned down by the Secretary of Defense for reasons of economy, but the work at *Redstone* Arsenal continued anyway, using some strange kind of book-keeping known only to Dr. von Braun and his associates!

On 4 October 1957, Radio Moscow announced that the Soviets had placed a satellite in orbit 900 kilometres above the earth. It was a sphere 58 centimetres in diameter, weighing 84 kilograms, nearly half of which was the weight of the batteries that powered its radio transmitter for 23 days, and it had four 'whip' antennas 1.5–3 metres long. Its apogee, or highest point in orbit, was indeed more than 900 kilometres above the earth; its perigee, or lowest point, was about 227 kilometres. The Soviets called it *Sputnik I* and said that it contained little in the way of instrumentation and was intended only as a test vehicle. However, the impact of this event was profound. While everyone connected with the American programme had been talking his head off and had nothing yet to show for it, the Soviets had said very little until their *Sputnik* was actually in orbit. Clearly, they had used their large intercontinental ballistic missile launcher, with just one additional rocket stage, to put their satellite into orbit, even though the first successful test of this missile had been announced only six weeks earlier. The military significance of this fact did not by any means lessen its impact. Furthermore, the empty second-stage rocket which had also gone into orbit was big—big enough to be seen with binoculars or even the naked eye just after sunset or before sunrise when the sky was dark and the empty rocket shell was in the sunlight. One month later, on 3 November, the Soviet Union launched a second satellite, this time a really big one, with about 500 kilograms of payload, including ten space physics experiments and a live dog named Laika. There was no provision for recovering the dog, of course, so she died when the satellite batteries ran out after about seven days in orbit. Telemetered physiological data, however, showed that she had survived the launching and had adapted reasonably well to the satellite environment.

Explorer 1 is launched

About a week after *Sputnik I*, a new American Secretary of Defense gave Dr. von Braun full authority to move his launching programme forward with top priority. The IGY committee assigned a cosmic-ray experiment by Professors James Van Allen and George Ludwig of Iowa State University to a small satellite designed by the Jet Propulsion Laboratory, California Institute of Technology, to be launched by the new *Redstone* launcher. On 31 January 1958, just 110 days later, this satellite, called *Explorer 1*, was put into orbit successfully. It weighed only 14 kilograms, including the empty fourth-stage rocket shell, to which it remained attached, and its payload was a miniscule 8·2 kilograms, but its orbit was higher than either of the two *Sputniks* that had already been launched. This fact, plus considerable genius on the part of Professor Van Allen and his associates, made possible a very important discovery—the existence of energetic charged particles trapped in the earth's magnetic field, now often called Van Allen radiation. This discovery was confirmed shortly thereafter by similar instruments on *Explorer 3*, which also contained a tiny tape-recorder capable of storing information over parts of the orbit when the satellite was not within range of a ground station.

The original Navy launcher system, called *Vanguard*, eventually began to function successfully and was used to orbit a grapefruit-sized test satellite weighing only a few kilograms. On 17 March 1958 this unpretentious little satellite went into the highest orbit yet; it was almost negligibly affected by atmospheric drag and its transmitter was the first to be powered by solar cells, so it continued to transmit for about six years instead of only a few weeks. Long-term tracking of this object later produced the first new information about the shape of the earth. However, it was not until 1959 that the unlucky *Vanguard* programme finally achieved a successful launching of the 10 kilogram scientific satellite for which it was designed.

That is how it all began.

2 Overcoming gravity

Everything that goes up
Must come down.
Anonymous

The well-known expression above must have been applied, at one time or another, to practically everything, including aircraft and the stock market. Until the advent of space technology it has been, in the common experience, universally true. But now we find that it is possible for something to go up and not come down again. How is it possible for a man-made object to be sent into orbit and travel completely around the earth, not just once but a thousand or a hundred thousand times, without any further expenditure of energy? Although the reader of today, having been educated by television and newspapers, will not find this question nearly so disturbing as he might have 25 years ago, it is useful to consider here how a satellite manages to stay in orbit, instead of inevitably crashing back on earth, as predicted by our opening cliche.

What keeps it up?

The answer is of course speed, or, more properly, velocity—since the speed must be appropriately directed. Consider the simple case of a man throwing a ball, as shown in Fig. 2.1. Let us suppose that he is standing on a small platform so that he releases the ball at a height of 4·9 metres above a flat playing field. If he should simply drop the ball it would accelerate downward under the influence of gravity at about 9·8 m/s^2,† reaching the ground in one second. If he should throw it horizontally at a speed of 14·7 m/s, it would still accelerate downward at the same rate, but it would also move away from him on a curved (parabolic) path, striking the ground at a distance of 14·7 metres. If he were able to throw it harder, say at a speed of 29·4 m/s, the path would have less curvature and the ball would strike the ground farther away; however, the downward acceleration would still persist at the rate of 9·8 m/s^2. Of course the atmosphere will resist any such motion, especially as the speed increases; but to avoid unnecessary complication we shall overlook atmospheric resistance at this point in the discussion.

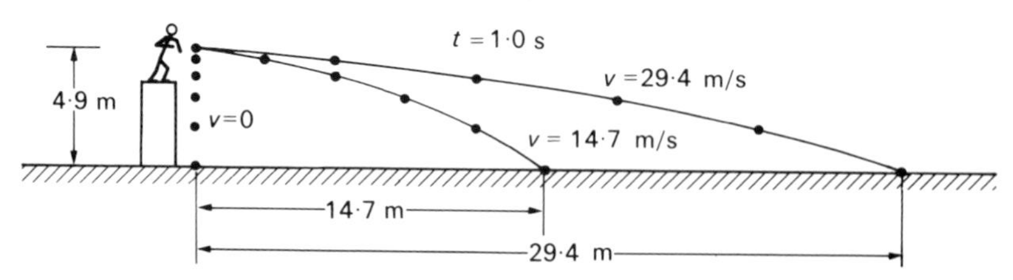

Fig. 2.1. Trajectories of a thrown ball over a flat earth.

In this simple example the earth was shown as a flat plane with the acceleration of gravity everywhere directed downward. Recognizing that the earth is actually

† m is the short notation for metre, s for second; km stands for kilometre and rad for radian, an angular measure.

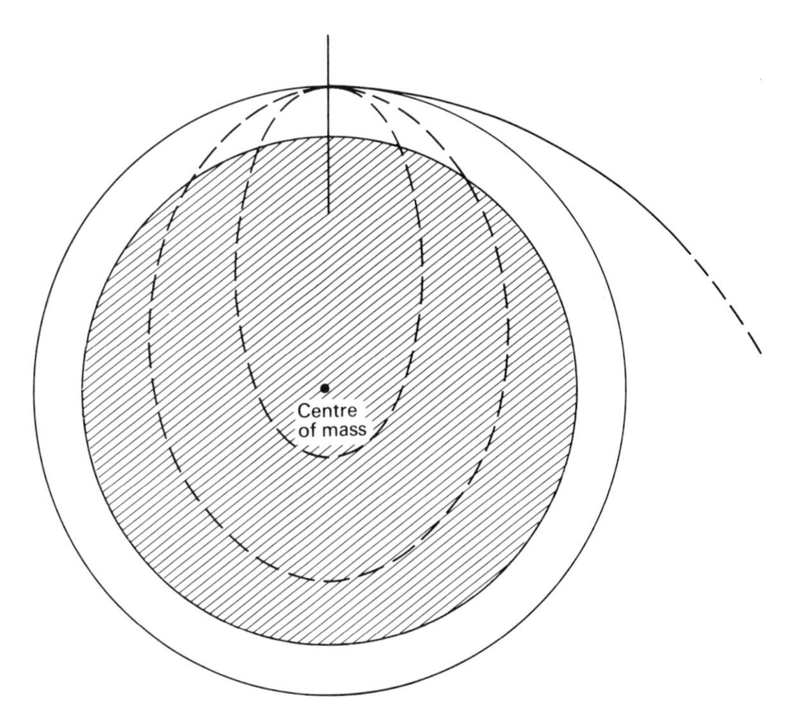

Fig. 2.2. Trajectories of a 'thrown' ball about a spherical earth.

spherical, rather than flat, with the acceleration of gravity everywhere directed towards its centre, we find that the parabolic paths of Fig. 2.1 become portions of ellipses, shown in Fig. 2.2. These ellipses are highly eccentric (thin) when the horizontal speed imparted to the ball is relatively small but become less and less eccentric (that is, fatter) as the speed is increased. Above some critical velocity, the trajectory ceases to intersect the earth's surface at all. If our ball player could throw the ball fast enough, and of course if there were no atmosphere to resist the motion of the ball, the ball would continue to circle the earth perpetually, although continually falling downward, or perhaps inward would now be a better word. The ball would be 'in orbit'.

The bothersome atmosphere can be avoided by going above it. At an altitude of about 500 kilometres, the density of the atmosphere is so small—approximately 10^{-12} kg/m^3 compared with about 1 kg/m^3 at sea level—that it produces almost negligible resistance even at very high speeds. (The expression 10^{-12} is mathematical shorthand for 1/1 000 000 000 000.)

Circular orbits

We now turn our attention to the question of just how fast it is necessary to 'throw' the ball in order to achieve a circular orbit. If the ball were not being accelerated—that is, if no forces were acting on it, not even gravity—it would continue to travel in a straight line at constant speed, and of course would fly off into space. For the ball to curve round on a circular path, it must be continually accelerated inward toward the centre of the earth by the pull of gravity. We know from simple experiments

Fig. 2.3. Swinging a bola.

with an arrangement like the Argentinian Gaucho's bola (shown in Fig. 2.3) that the inward acceleration of a ball moving on a circular path is equal to the square of the velocity divided by the radius of the orbit. For a satellite in a circular orbit, this acceleration must always be exactly equal to the acceleration of gravity at the orbital altitude. From this equality, it is possible to calculate the velocity required to maintain a circular orbit at any altitude.

Thus, for a circular orbit at sea level (which would be possible only if we could somehow do away with the atmosphere and any mountains that might stand in the way) the required velocity would be about 7·89 km/s, (or 4·89 statute miles/s for those who still think better in the good old English units). For orbits at higher altitudes, the velocity varies inversely as the square root of the radius. Of course, the

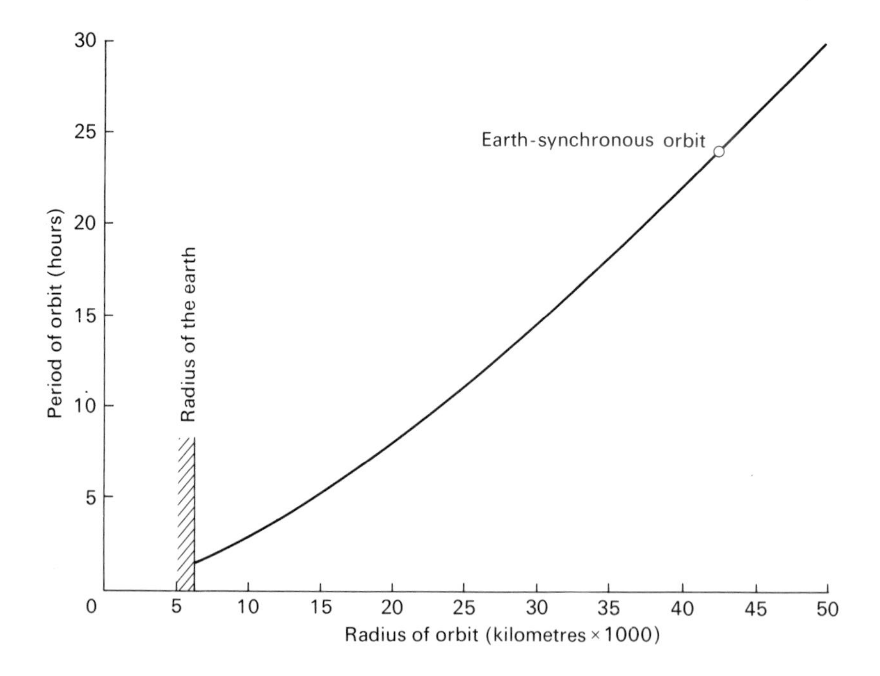

Fig. 2.4. Relationship between radius and period of earth-satellite orbits.

earth is not stationary but is rotating in such a way that a point on the equator will always have a velocity toward the east of 0·465 km/s. Thus if our ball player were to stand at the equator and throw the ball horizontally eastward, he would have to be able to throw it at a speed of about 7·43 km/s—a truly hypothetical capability.

At an altitude of 500 kilometres which, as we have noted, is high enough to avoid most of the effects of the atmosphere, the orbital velocity would be about 4 per cent smaller, or about 7·59 km/s (4·71 miles/s) (in round English units, this is approximately 25 000 ft/s). At this speed and altitude the angular velocity would be $1{\cdot}1 \times 10^{-3}$ rad/s, and the period, or time required, for a complete orbit with respect to fixed coordinates would be 5700 seconds or 95 minutes. For a sea-level orbit, if that were possible, the period would be about 84·5 minutes. The orbital period varies as the $\frac{3}{2}$ power or the radius, as shown in Fig. 2.4.

A particularly interesting orbit is the one for which the period is exactly 24 hours, because a satellite in such an orbit revolves about the earth in exactly the same length of time that it takes the earth to rotate on its own axis. This orbit is called an earth-synchronous or geosynchronous orbit. If a geosynchronous orbit lies in the plane of the earth's equator, the satellite in it will appear to an earthly observer to remain stationary while the sun, moon, and stars march past it in their daily progression across the sky. Such a satellite is therefore known as a geostationary satellite. If the geosynchronous orbit is inclined to the equator, the satellite will appear to trace out a figure of eight once each day. An observer or an optical instrument such as a telescopic camera in a geostationary satellite would always see the same portion of the earth's surface with the dawn, daylight, twilight, and night phases progressing across it every 24 hours. The advantages of the earth-synchronous orbit for such purposes as television broadcasting, radio-relay stations, and certain kinds of earth observations are obvious.

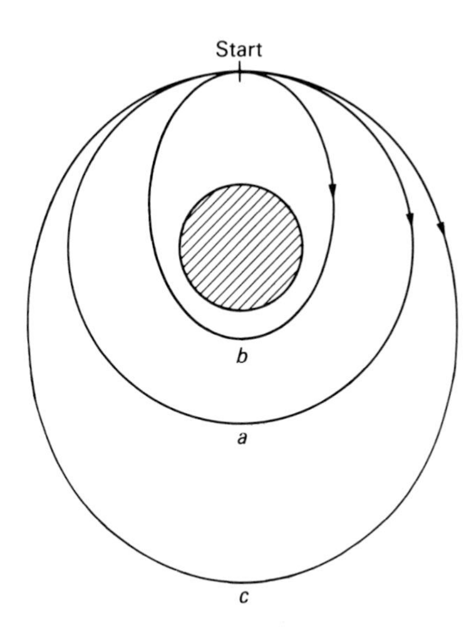

Fig. 2.5. Effect of initial speed of an orbit: *a*, circular; *b*, tilted up; *c*, tilted down.

Non-circular orbits

So far we have been considering only circular orbits. Nature puts no such constraint on what a satellite can do. If, for example, a vehicle at some point outside the earth's atmosphere be given a horizontal velocity of magnitude less than that required to establish it in a circular orbit, it will descend to a lower altitude, as in orbit *b* of Fig. 2.5. During the course of its descent the vehicle will acquire more speed than it needs to remain in a circular orbit at the lower altitude so it will rise again, losing speed as it does so, until it returns to the original conditions. Similarly, if the initial velocity is horizontal but more than required for a circular orbit, as in orbit *c* of Fig. 2.5, the vehicle will rise, losing speed for half an orbit, and then fall back to the starting point with the initial velocity. If the initial velocity is of the right magnitude

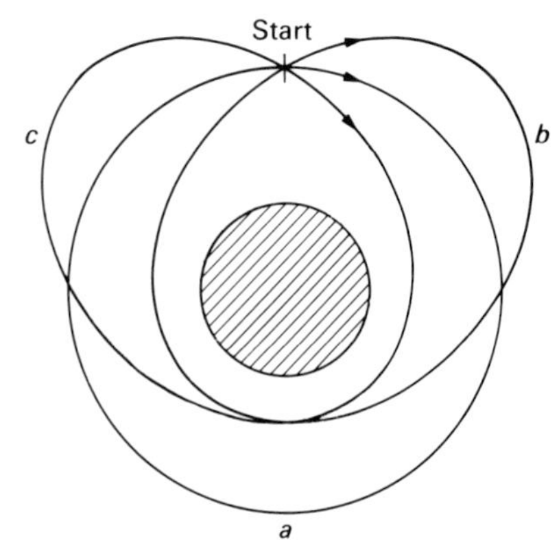

Fig. 2.6. Effect of non-horizontal direction: *a*, horizontal; *b*, tilted up; *c*, tilted down.

but directed above or below horizontal, as shown in Fig. 2.6, the vehicle will rise or fall until the radial component of its velocity becomes zero and reverses. If it does not encounter the earth or the denser parts of the earth's atmosphere, it will return at the end of each orbit to its starting point with the original velocity. The points on an orbit that are nearest or farthest from earth are called perigee and apogee, respectively, from the Greek *peri*, and *apo* meaning near and far, and *geos*, meaning earth. A mathematical treatment of satellite orbits in the Appendix (p. 167) shows that these non-circular orbits are ellipses. The Appendix also shows that the total energy of a satellite in any given orbit is constant. Once in orbit a satellite will remain there without any further addition of energy (propulsion). In an elliptic orbit some of the energy is continually being transformed from kinetic to potential energy and back again, but the total remains the same unless the orbit is changed. Somewhat surprisingly, it is found that orbital energy depends only on the length of the major axis of an elliptical orbit and not at all on the length of the minor axis, which determines the 'thinness' or 'fatness' of the ellipse. Thus orbits *b* and *c* in Fig. 2.6 have major axes equal to the diameter of orbit *a*, because all three start out with the same total energy.

For a circular orbit at an altitude of 500 kilometres the energy per unit mass would be $3{\cdot}34 \times 10^7$ J/kg (joules per kilogram). In terms more familiar to those who pay the electricity bill each month, this is a little less than 10 kilowatt hours per kilogram of satellite and payload. If it were all supplied as an impulse in the form of kinetic energy, it would correspond to a velocity of 8·18 km/s. The energy thus calculated could be considered to be the energy required to launch a satellite if theoretically ideal processes could be used for the launching. Actually the processes used are far from ideal (in this sense), so more energy is needed.

Changing the orbit

If we have already placed a satellite in a low circular orbit around the earth and we wish to increase the height of the orbit, for example, to that of an earth-synchronous orbit, a convenient way to do so is first to increase its energy by an application of thrust so as to change it to an elliptical orbit with an apogee at the desired new altitude, and then to increase the energy again by applying a thrust at apogee so as to make the orbit circular once more. This technique, which is illustrated in Fig. 2.7, is popularly known by American engineers as the 'kick in the apogee' technique.

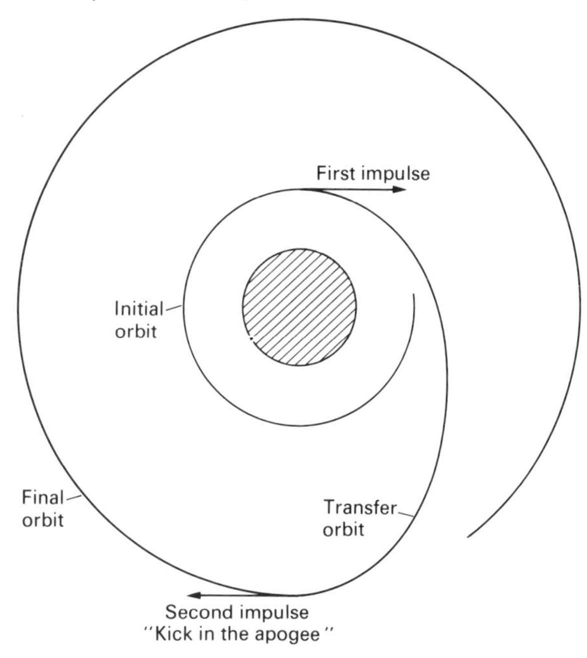

Fig. 2.7. Enlarging an orbit using a transfer ellipse.

The influence of atmospheric drag on a satellite orbit is in a way almost the converse of the thrust at apogee just described. The density of the atmosphere increases almost logarithmically as the satellite approaches the earth and the satellite velocity also increases. Thus most of the drag occurs near perigee and can be thought of as repeated negative energy increments at this point in the orbit. The effect, as shown in Fig. 2.8, is not to decrease the height of the perigee, where the drag occurs, but rather to decrease the major axis of the orbit without changing the perigee. The net effect is a gradual reduction in the apogee until the orbit becomes almost circular. The satellite then spirals in toward the earth, gaining velocity and therefore kinetic energy, but losing potential energy at twice the rate at which it gains kinetic energy,

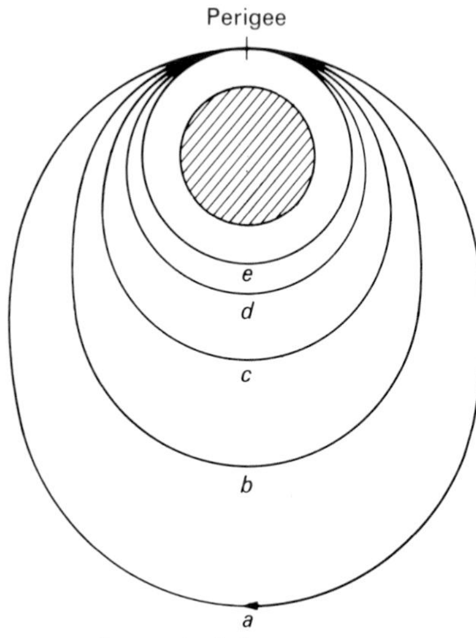

Fig. 2.8. Effect of atmospheric drag on an elliptic orbit.

so that its total energy continues to decrease, as it must to account for the energy dissipated in the atmosphere and in heating the satellite.

Actually, the earth does not have perfect spherical symmetry as has been assumed in all of the foregoing. Its most significant departure from that ideal figure is similar to that of a middle-aged man—a slight bulge around the middle. Its equatorial radius is about 21·5 kilometres greater than its polar radius. It will not be possible in this book to examine rigorously the effects of such a mass distribution on a satellite orbit, but a brief qualitative discussion of one of these effects seems desirable.

A greatly exaggerated bulge is shown in Fig. 2.9. The net effect of this bulge on a satellite in an inclined orbit is to add a small non-radial component of acceleration in a southerly direction while the satellite is over the northern hemisphere and in a northerly direction while it is over the southern hemisphere. These perturbing accelerations cause the orbit to precess like a toy gyroscope does when one applies a torque perpendicular to its spin axis. Anyone who has ever tried this little experiment will recall that instead of obediently turning in the direction one wants it to the gyroscope perversely insists on turning about an axis perpendicular to that of the applied torque. Thus the effect of the bulge is not to decrease the angle of inclination of the orbit, as might naively be expected, but rather to cause the orbit to precess about the earth's axis of symmetry (polar axis).

If the motion of the satellite round the earth is generally eastward, as shown in Fig. 2.9, the direction of the precession will be westward, or clockwise as seen looking down on the North Pole. However, this direction could be reversed by causing the satellite to move in the opposite direction. Such an orbit would be called a retrograde orbit because the satellite motion would then be in the opposite direction to the motion of the earth about its own axis. For a retrograde orbit the precession would therefore be counter-clockwise as viewed from above the North Pole, which is the same direction as the motion of a line from the earth to the sun. It turns out

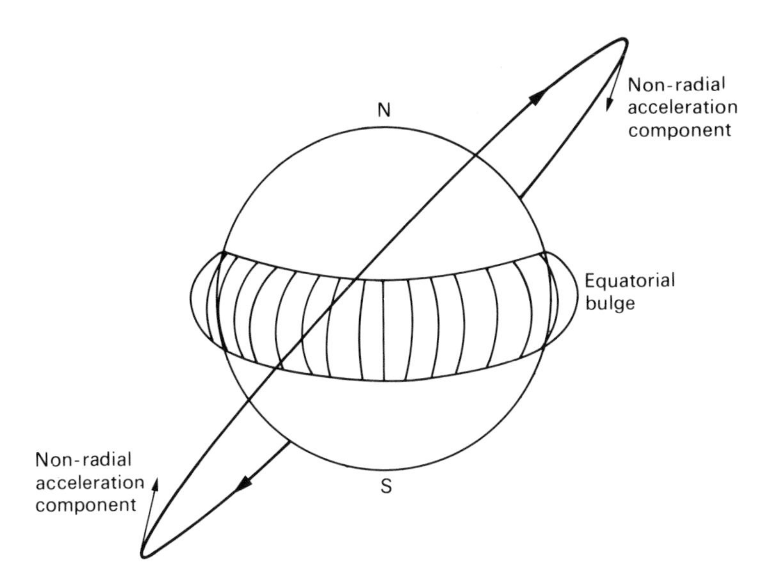

Fig. 2.9. Gravitational effect of the earth's equatorial bulge on a satellite orbit.

that these two motions can be made exactly the same by putting the satellite in a nearly circular orbit having an inclination of about 99·1° (80·9° retrograde) at about 910 kilometres altitude. This is called a sun-synchronous orbit because its orbital plane is fixed relative to the direction of the sun—earth line and therefore the satellite always passes over each point on the earth at the same local or 'sun' time. Such an orbit has a significant advantage for certain kinds of earth-observation experiments and applications.

Rocket launchers

The question must now be asked, how can the necessary kinetic and potential energy be given to a vehicle in order for it to be inserted into the desired orbit? Clearly this cannot be accomplished by firing the vehicle out of a gun barrel, no matter how big the barrel or how long. Even if the requisite energy could be transferred to the vehicle, which does not appear to be feasible with any conceivable gun technology, it should be apparent from the foregoing discussion that any orbit originating at the surface of the earth must intersect the surface of the earth and therefore be of only hypothetical interest! The only answer to this question so far has been the use of rocket propulsion, although it is not inconceivable that a system might be devised in which a part of the energy would be supplied by a very large jet aircraft used to carry a rocket launching vehicle to supersonic speed high in the stratosphere.

Rocket propulsion is a form of jet propulsion which differs from the 'air-breathing' form used in modern aircraft engines in that all of the propellant mass which is ejected at high speed to form the jet is carried in the vehicle at take-off. In an aircraft jet engine, external air is drawn in, compressed, heated by the combustion of fuel with some of the oxygen in the air, and finally expelled through a nozzle. In a rocket, fuel and oxidizer contained within the vehicle are burned together in a combustion chamber, and the hot combustion gases are expelled through a nozzle as in

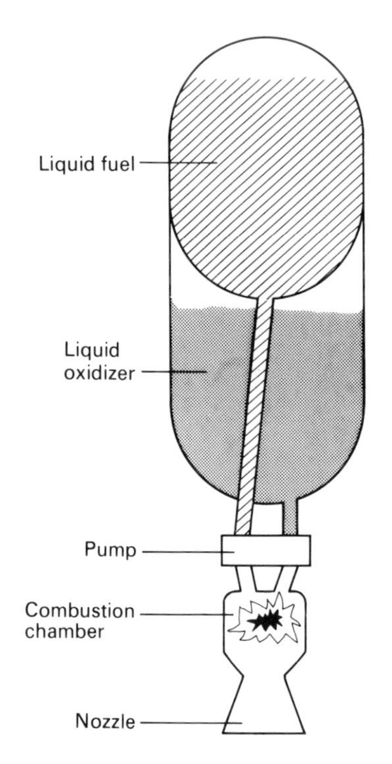

Fig. 2.10. Liquid-propellant rocket.

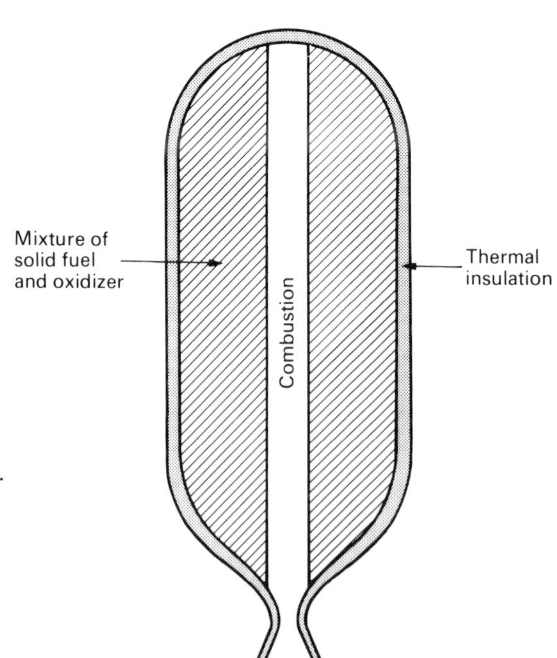

Fig. 2.11. Solid-propellant rocket.

the aircraft engine. Liquid-fuel rockets use a liquid fuel and a liquid oxidizer carried in separate tanks, which are pumped or forced by gas pressure into a small combustion chamber to which the nozzle is attached (Fig. 2.10). Solid-fuel rockets use an intimate mixture of solid fuel and oxidizer loaded into a very large combustion chamber with a nozzle on the end (Fig. 2.11). In either case, the thrust results from the reaction of the hot gas expanded through the nozzle.

How a rocket motor works

In order to understand how a rocket motor 'works' consider the hypothetical situation shown in Fig. 2.12 in which a demon inside the rocket throws out little balls, one after the other, at a high velocity with respect to the rocket. As each ball is thrown, its momentum is changed. If we consider momentum to be positive to the

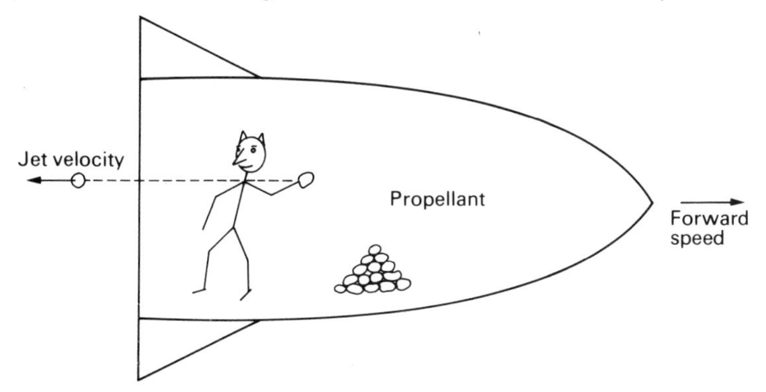

Fig. 2.12. Principle of rocket propulsion.

right in this diagram, the change in momentum of the ball will be negative. If there are no external forces such as atmospheric drag acting on the closed system consisting of the rocket and its contents including the little balls (propellant), its total momentum must remain constant. Therefore, after a ball has been thrown the momentum of the rocket and contents, now less by one ball, must be slightly greater in the positive direction (that is, to the right) than before. As one ball after another is thrown to the left, the rocket and its remaining contents are made to go faster and faster or, as one would say, are propelled to the right. The physical explanation for this phenomenon, as every cricket or baseball player knows, is that in order to throw the ball to the left, the little demon must push with his feet to the right.

Note that the process we have been describing does not depend on there being any air to 'push against', nor does it make any difference how fast the rocket is moving, even if it is moving at a higher velocity than the demon can throw the balls. However, it is important for him to throw the balls as hard as he can because the higher their velocity with respect to the rocket, the greater will be their change in momentum and the greater will be the propulsive effect each throw will have on the rocket.

It is a curious fact about rocket propulsion that if the rocket starts from zero speed, in the absence of gravity or atmospheric drag, the centre of gravity of the whole system including the balls that have been thrown would go nowhere! It would simply remain stationary, the rocket itself moving off to the right exactly balanced by the little balls or propellant speeding off to the left.

In a real rocket, of course, the busy little demon must be replaced by a pressure chamber and nozzle and the little balls by molecules of hot gas; however, the principles are the same. The so-called thrust or force exerted by a rocket engine is equal to the rate of change of momentum of the propellant gas as it is accelerated through the nozzle. This can be calculated by multiplying the mass flow rate of propellant by the velocity, relative to the rocket, at which it is expelled. A rocket engine is usually rated not in terms of efficiency but rather in terms of the amount of impulse (thrust muliplied by time) it can deliver per unit weight ot propellant expended. This quantity is equal to the jet velocity divided by the sea-level acceleration of gravity and has the units of time (seconds).

A 'good' rocket engine, therefore, is one that has a high jet velocity. Going back to the little demon for a moment, it is obvious that he can throw the balls faster if he is strong and if the balls are not too heavy. Similarly, high jet velocity is obtained by using high-energy propellants with products of combustion that have low molecular weight and by designing the rocket to operate at high pressure. A well-designed rocket engine using liquid hydrogen and liquid oxygen as propellants typically produces a jet velocity of about 3500 m/s, which is equivalent to a specific impulse of 357 seconds.

Even though the thrust of a rocket is held constant, its velocity increases faster toward the end of burning than at the beginning because, having expended most of its propellant, it is lighter. Using a little maths we could show that if a rocket vehicle were to travel in a straight line in a gravitationless vacuum the increase in its velocity from the beginning of burning until the end would be equal to the jet velocity multiplied by the natural logarithm of the ratio m_1/m_2, where m_1 is the mass of the rocket and its contents at the start and m_2 at the end of burning. For a mass ratio greater than 2·718, which is by no means impossible, the vehicle would acquire a forward velocity greater than the relative velocity of the jet with respect to the rocket. Under these conditions, the gas in the jet as well as the rocket itself would be moving forward rather than backward as seen by a stationary observer, but this seemingly disconcerting state of affairs would have no effect whatever on the thrust.

Actually, it is quite possible with modern design practice to make the 'dead' weight of a large rocket—that is, the weight of everything except the payload and propellants—less than one-tenth the weight of the propellants, so that if only a negligible payload were to be carried, the logarithm of m_1/m_2 would be about 2·4 and if the exhaust-gas velocity were 3500 m/s the rocket velocity at burn out would be 8·4 km/s. Thus a liquid-hydrogen/liquid-oxygen rocket of reasonable design, carrying only a small paylord, could, at least marginally, be expected to achieve orbit.

Multi-stage rockets

However, it is rather obvious that the purpose of a rocket launcher is to put as much payload into orbit as possible, and higher velocity is required for earth-synchronous orbits, exploration of the moon, and other such interesting purposes. It seems fairly clear that a simple rocket such as we have been describing cannot accomplish all that might be desired, no matter how large it is made. So early rocket pioneers came up with the idea of multiple-'staged' rockets, which to a certain extent overcame this limitation. The basic idea is to use one rocket with its propellant as the payload of another larger rocket and if necessary make this larger rocket the payload of a still larger rocket, and so on to as many stages as required. The advantage of this scheme is that the empty tanks, engine, and structure of the larger rocket stages can be discarded at relatively low altitude and velocity, instead of being carried completely into orbit as in the case of a single-stage rocket.

As an example, consider a two-stage rocket, using liquid hydrogen and liquid oxygen as propellants, and having a ratio of dead weight to propellant of one to ten in both stages. Let it also be assumed that the fully loaded weight of the second stage is equal to twice the dead weight of the first stage and the payload equal to twice the dead weight of the second stage.

For this hypothetical rocket, the mass ratio of each stage would be 13/3 or 4·33, of which the logarithm is 1·465. Thus the total gravitationless vacuum velocity would be about 10·2 km/s, enough to reach an orbit well above 500 kilometres altitude. The initial mass of the second stage would be 2/13 the initial mass of the first stage, and the mass of the payload 2/13 the initial mass of the second stage. Thus the payload would be $(2/13)^2$ or about 1/42 of the initial take-off mass. In other words it would require 42 tonnes take-off weight to put 1 tonne of payload (plus the $\frac{1}{2}$ tonne dead weight of the second stage) into an orbit that could not have been achieved, even without any payload, by a single-stage rocket using the same propellants and similar construction techniques. The advantage is rather obvious! So obvious, in fact, that it is surprising to note that the first experiment with a liquid-fuel two-stage rocket was made only after the Second World War, using a modified German V-2 rocket as the first stage and a small research rocket designed by the California Institute of Technology as the second stage. This combination was known as *Bumper*, and it worked!

Modern rocket launchers generally consist of two, three, or four stages, and sometimes use equivalent techniques such as 'strap-on' solid-propellant rocket boosters that burn only a short time and then are discarded, or propellant tanks that can be emptied and then dropped off early in the propulsion phase of the flight. These launchers, which vary in size from the 68-foot *Scout* to the giant *Saturn* and similarly large Soviet rocket vehicles, will probably be superseded some day by recoverable or partially recoverable launchers such as the *Shuttle*, about which more will be said in Chapter 11.

Thus, it is clear that although gravity itself has not been overcome in the sense of being eliminated or abolished, the limitations which it was once believed to impose on man's ability to send himself and his instruments far above the earth's atmosphere have indeed been overcome—by the powerful thrust of rockets.

3 Designing for the satellite environment

> . . . *and the work some praise and some the Architect; his hand was known In Heav'n by many a Tow'red structure high, Where Scepter'd Angels held thir residence,*
>
> John Milton (1667) *Paradise lost*, Book I

An artificial earth satellite is not exactly a 'tow'red structure', and it seems unlikely that many astronauts would qualify as 'scepter'd angels'. However, the design of a satellite to operate in the heavenly environment is by no means an easy task; the 'architect' who can design a successful one certainly deserves praise.

In the first place, because it costs many thousands of dollars per kilogram to launch a satellite, its cost (in orbit) is greater than the cost of an equal mass of pure gold. Consequently the designer must be something of a genius at the task of minimizing weight. Superfluous 'safety factors' which in reality are often only factors of ignorance cannot be tolerated and only the strongest, but lightest materials can be used.

In the second place, because there is no possibility of sending a service man around with a truck load of spare parts to repair a satellite if some critical subsystem should fail in orbit—or at least there won't be any such possibility until the new *Shuttle* becomes operational by around 1980 (see Chapter 11)—the designer must strive for the ultimate in reliability, using every technique at his disposal toward this end. These techniques include simplicity of design, extensive testing of each material, piece, and component, judicious employment of redundance, thorough systems analysis to determine the influence of each part on every other, and finally a comprehensive knowledge and understanding of the special environment in which the satellite must operate. It is true that the presence of clever astronauts in the satellite, in communication with teams of experts on the ground, can sometimes make it possible to repair an unexpected failure in orbit. However, the presence of human beings in the satellite requires even more emphasis on design reliability. Loss of human life in a space programme is not something that can be tolerated by society.

A vacuum to travel in

Of the environmental factors that must be taken into account by a spacecraft designer, the one that comes most quickly to mind is the nearly perfect vacuum or lack of atmosphere which prevails at an altitude of more than a few hundred kilometres above the earth's surface. It has been noted in the previous chapter that lack of atmospheric drag is one of the essential elements required to maintain a satellite in orbit. However, this same lack of atmosphere also represents a problem to be dealt with in designing the satellite.

Providing an environment for the spaceman

If there are to be live people or animals on board, the space in which they are to be enclosed must contain a suitable atmosphere. The normal atmospheric environment near the earth's surface consists of about three-quarters nitrogen, a little less than one-quarter oxygen, and small amounts of carbon dioxide, water vapour, and other gases. The sea-level pressure of this mixture is about 10^5 N/m^2 (10^6 dyn/cm^2).† One alternative open to the designer, in fact the one chosen by Soviet spacecraft designers, is to provide approximately this same atmosphere in an enclosure capable of withstanding full atmospheric pressure. Another alternative is to reduce the total pressure while increasing the fractional oxygen content so as to maintain nearly the same partial pressure of oxygen. This second alternative, which has been used in manned spacecraft of the United States, saves considerable weight in the pressure enclosure and has been found to be physiologically acceptable, but it does result in an atmosphere with high oxygen content and very little inert gas as diluent. Many materials not ordinarily thought of as particularly combustible will burn vigorously in such as atmosphere, and it was this phenomenon that resulted in the tragic fire during a pre-launch test of an *Apollo* spacecraft in the United States. Any failure of the pressure enclosure, of course, means almost instant death for the persons inside, because the fluid in their bodies will immediately begin to change to vapour—quite literally to boil—at a pressure less than about one-hundredth normal atmospheric pressure.

Unmanned spacecraft do not necessarily need to be pressurized. An advantage of pressurization is that the internal atmosphere can be used for distributing the internally generated heat, that is, to cool certain components that might otherwise overheat. However, with careful thermal design, components can be cooled by conduction through metallic surfaces, by radiation to other cooler surfaces, or by a relatively new technique known as the 'heat pipe'. This latter makes use of a fluid which evaporates at the hot end and condenses at the cool end of a small hermetically sealed tube with a wick or some other means to return the condensed fluid to the hot end. Most American designers conclude that the disadvantages of weight and possible failure of the enclosure for a pressurized (unmanned) spacecraft outweigh the advantage in thermal design.

Insulation and lubrication

There is also the question of electrical insulation. Air, or almost any other gas, at sea-level pressure is a reasonably good insulator, able to sustain an electrical potential gradient on the order of 3×10^6 V/m‡ without breakdown. A nearly complete vacuum also has good insulating characteristics. At intermediate pressures, however, air breaks down relatively easily. Consequently, if the spacecraft is not hermetically sealed and pressurized it must be designed with large vents so that the atmosphere which it contains at launch can escape quickly and high-voltage circuits must not be energized until after complete evacuation. Leakage from pneumatic systems and evolution of gas by evaporation of plastics or lubricants must also be avoided.

The design of sliding or rubbing surfaces such as bearings for operation in a vacuum poses a special problem. Liquid or paste lubricants will evaporate rapidly unless sealed into the bearing. A reservoir of lubricant can be provided but it is very difficult to make a seal with small enough leakage to avoid contamination of other space-

† N represents newton, the unit of force. A dyne (abbreviated dyn) is the unit of force in the c.g.s. system.

‡ 3 million volts per metre.

craft components. The oxide surfaces and the layers of adsorbed gas which normally protect metallic parts in air do not form again when removed by abrasion in a vacuum. Thus dry rubbing surfaces tend to seize or vacuum-weld. There are some dry lubricant materials which show promise, an example being molybdenum disulphide. These materials cling tenaciously to metallic surfaces, yet produce low friction when rubbed against themselves.

The absence of gravity

Another aspect of the satellite environment which comes readily to mind is the absence of gravity, the condition known to space technologists as 'zero-*g*'. In a strict sense, of course, gravity is not absent. As noted in the previous chapter, it is the attraction of gravity which constrains the satellite to revolve about the earth instead of flying off into space. However, the motion of the satellite is such that the gravitational field at its centre of gravity is exactly balanced by an equal and opposite acceleration field. The situation is similar to that which exists in a freely falling elevator cage, except that it does not end in a crash at the bottom of the shaft! An object released with zero relative velocity inside the satellite, or outside it for that matter, will float along in the same relative position without any support, just as it would if there were indeed no gravity. This is an environmental condition that is essentially unique to the space environment. It cannot be simulated on earth except for a few seconds of free fall in a tall evacuated drop tower, or for about a minute in an aircraft flying on a ballistic trajectory.

Of course any externally applied force such as atmospheric drag or the firing of a rocket thruster will cause an acceleration of the satellite that is not shared by objects inside until they come in contact with the satellite structure. And if the satellite is caused to spin, all parts of the spacecraft and its contents will be accelerated away from the spin axis at a rate proportional to the distance from the axis multiplied by the angular velocity squared. If an artificial gravitational field should be required, the spacecraft could be designed in the shape of a huge dumbbell or doughnut rotating about a transverse axis, with the payload located near the radial extremities. It does not now seem likely that artificial gravity will be needed even for manned flights of long duration. Some spacecraft, especially the smaller ones, are designed to spin, but this is done to provide a simple means of attitude stabilization rather than artificial gravity.

Stabilization

It should be noted that the gravitational field and the acceleration field cannot be exactly equal except at the centre of mass of the satellite, because the gravitational field varies as the inverse square of the distance from the earth's centre of mass and some parts of the satellite are at a greater distance than others. This is an extremely small effect. For the hypothetical dumbbell-shaped satellite shown in Fig. 3.1, consisting of two masses connected by a stiff weightless rod oriented toward the centre of the earth, the part of the satellite farthest from earth would have an apparent acceleration with respect to the satellite centre of mass directed radially outward, and the part nearest earth would have a similar acceleration directed inward. If the separation between the two parts were 20 metres and if the whole satellite were in an orbit at 500 kilometres altitude, the magnitude of this apparent acceleration

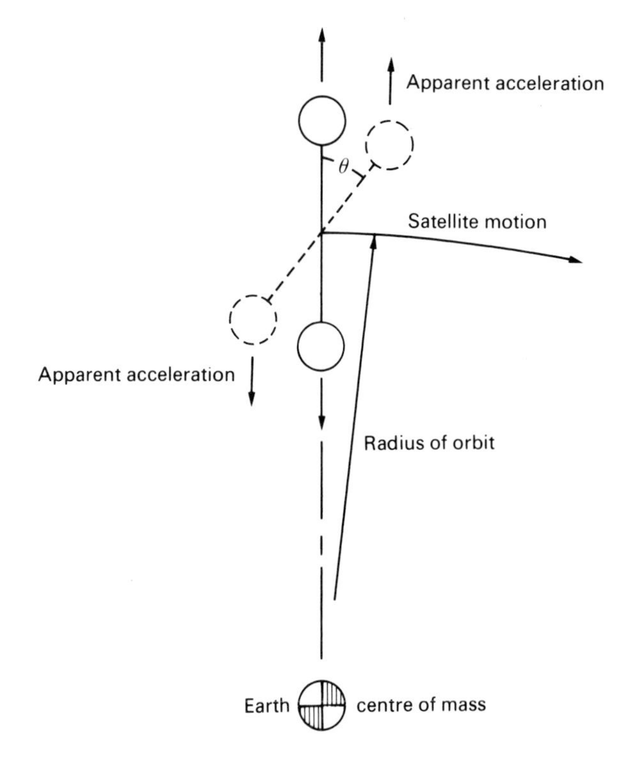

Fig. 3.1. Apparent accelerations used for gravity-gradient stabilization.

would be about $2{\cdot}5 \times 10^{-6}$ g (1/400 000 times g, the acceleration of gravity at sea level). However, even this miniscule acceleration can be detected and can be used to stabilize the satellite in an earthward-pointing direction. If the satellite should be displaced from the radial alignment by an angle θ, as shown in dashed lines in Fig. 3.1, there would be a restoring torque proportional to θ for small displacements. This would cause the satellite to oscillate about its radial alignment with a period of about 67 minutes, a value which is independent both of the masses of the two parts of the satellite and of the length of the connecting rod. The oscillation could be made to die out by means of appropriate mechanical or electromechanical damping devices, so that the satellite would be stabilized in the desired mode. This method of stabilization, known as gravity-gradient stabilization, has been used successfully on satellites in relatively low, nearly circular orbits. For obvious reasons it becomes much more difficult to apply if the orbit is very high or very elliptical. Even for ideal orbits its accuracy is limited by the existence of anomolies in the earth's gravitational field and other sources of 'noise', so it cannot be considered for applications that require extremely precise stabilization. Where it can be used gravity-gradient stabilization has the advantage of simplicity and long life.

Problems with fluids and gases

The zero-g environment influences design in many ways that are not immediately apparent. For example, a fluid does not accumulate in the bottom of a tank, where it can be conveniently pumped out or driven out by gas pressure. Instead, if it can wet the inner surface of the tank most of the fluid will cling to the tank walls leaving an irregular void near the centre, and if it cannot wet the walls the fluid will accumulate in balls or globules anywhere in the tank. In order for the fluid to be moved from one container to another it must be confined in a flexible bladder with no gas bubbles. In general, liquids must be moved by pumping, by squeezing, or by wicking, which depends on capillary forces; they cannot be poured or drained.

Thermal convection will not operate in zero gravity. A candle will not burn, even in an atmosphere containing plenty of oxygen; it quickly surrounds itself with gaseous combustion products and goes out. Oxygen can reach the flame only by diffusion, which is too slow. A breathing animal does better than the candle because it can create some circulation and mixing by the force of its exhalation and by moving about. For heat removal from satellite components, however, forced circulation must be provided by means of a fan or blower.

Building in space

From a structural point of view, the lack of gravity can be an advantage. Of course ground-assembled satellite structures must be designed for the high accelerations of launching. However, it is believed that very large radio-telescope reflectors could be assembled in space with great geometric prevision using light-weight structural elements. In fact, there seems to be no technological limit to the size and complexity of structures that could be assembled in this way. A complete space hotel, for example, has even been designed. Perhaps the 'many tow'red structures' envisaged by Milton will yet be realized!

Upper atmosphere and ionosphere

Although the earth's atmosphere at satellite altitudes is a very good vacuum it is not an absolute void. At 500 kilometres its density varies from about 10^{-12} kg/m^3 during sunspot maximum to about 10^{-13} kg/m^3 during sunspot minimum (see Fig. 3.2). The temperature of such a thin atmosphere cannot be measured by an ordinary thermometer; the temperature of any such device would be controlled primarily by radiative equilibrium between the device and the cold black sky, the warm earth, and the hot sun. However, temperature still does have its usual significance as a statistical measure of the random motion of the atomic and molecular particles which constitute the atmosphere, and the laws of thermodynamics still apply if the dimensions of the sample be made large enough. At this altitude the atmosphere is essentially isothermal—that is, its temperature does not change with altitude, although it does drop from about 1600 K at sunspot maximum down to about 900 K at sunspot minimum. (The values 1600 K and 900 K refer to degrees on the Kelvin temperature scale, which is the same as degrees centigrade plus 273.) This variation with solar activity results from absorption by the upper atmosphere of greater amounts of ultraviolet, X-ray, and hydromagnetic wave energy from the sun during periods of enhanced solar activity. There is, of course, a diurnal variation of temperature and density in the upper atmosphere and also a day-to-day variation that seems to be correlated with individual solar disturbances.

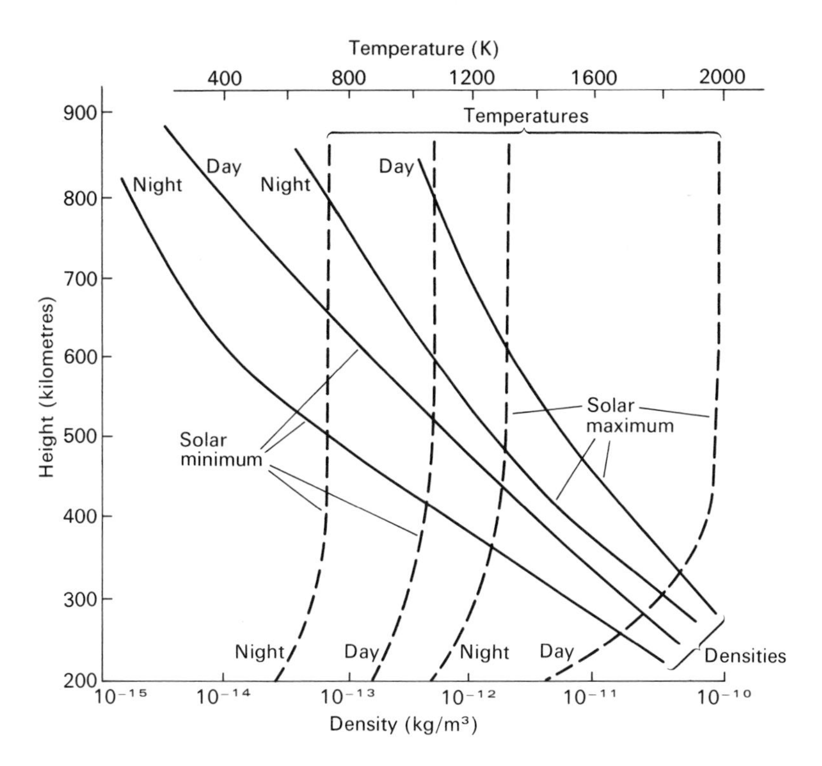

Fig. 3.2. Average atmospheric density and temperature.

The foregoing information is significant to the satellite designer primarily when he is concerned with making a satellite to fly at the lowest possible altitude or with the smallest possible perturbations from a precisely calculated orbit. Actually, much of the information we now have about the upper atmosphere has been obtained by precise measurements from the ground, over a long period of time, of just such small perturbations in satellite orbits.

The ultraviolet and soft X-radiation from the sun is highly variable with solar activity, but always has sufficient energy to cause significant photo-ionization of the earth's atmosphere at high altitudes. Recombination of the electrons and ions thus produced is so slow at these altitudes that substantial concentrations of electrons can persist even throughout the night. Cosmic rays and energetic particles from the sun also contribute to some extent to the continued state of ionization of the upper atmosphere. This partially ionized region is appropriately called the ionosphere.

Problems of communication

The presence of free, low-energy electrons in the ionosphere makes it electrically conductive, at least for low-frequency currents. Thus, fortunately, a spacecraft orbiting within the ionosphere cannot accumulate large surface charges that might result in high potential differences between insulated areas on its outside surface. The free electrons in the ionosphere also increase the relative permittivity (dielectric constant) of the atmosphere, causing radio waves to travel more slowly in regions where the electron density is greater and to be reflected or refracted (bent) by the boundaries of such regions. In the lower part of the ionosphere the electrons can also absorb some of the energy of radio waves. Both of these effects depend on the frequency of the transmission. In general, radio communication between the satellite and ground, which must pass through the ionosphere, cannot be accomplished using low- or medium-frequency transmissions, and is unreliable or 'freaky' at high frequencies. For good communication on a satellite-ground link it is necessary to use a frequency of at least 50 megahertz (50 million cycles per second).

Whistlers

The very highest level of the atmosphere, if indeed it can still be called atmosphere, extends from 1100 kilometres or 1500 kilometres outward and consists of approximately equal concentrations of protons (atomic hydrogen ions) and electrons. This protonosphere as it is sometimes called is the medium responsible for propagation of radio 'whistlers'—low-frequency (1–30 kilohertz) electromagnetic waves that follow paths along the earth's magnetic field lines from one hemisphere to the other. They are believed to be initiated by lightning strokes near the earth's surface. Because of the dispersion (variation of propagation speed with frequency) of the transmission path, the impulsive signal of the lightning stroke is 'stretched' out into a continuously varying tone or whistle. Hence the name. Until the use of high-altitude rockets and satellites became possible, the main experimental evidence about electron densities at great heights was obtained through ground observation of the characteristics of these waves.

The thermal environment

As previously noted, the temperature of the atmosphere at typical satellite altitudes is on the order of 1500 K. Furthermore, the satellite moves at such a high speed that the atmospheric particles with which it collides would appear to an observer in the satellite to have a much higher average energy than to a stationary observer in the atmosphere. However, none of this has any significance for the engineer who is responsible for the thermal design of a satellite because the atmosphere is so diffuse that it is unable to transfer much heat energy to the satellite regardless of its temperature. To get an order-of-magnitude feeling for the validity of this statement, recall that the density of the atmosphere is something like 10^{-12} kg/m^3 at 500 kilometres altitude, and consider a satellite one square metre in cross-section moving through this atmosphere at 10 km/s. Under these conditions, it would physically interact (collide) with about 10^{-5} grams of atmosphere per second. This is approximately the mass of a dust particle 100 micrometres in diameter—smaller than the full stop that ends this sentence. Such a small amount of mass, even though it be at very high temperature and impacting the spacecraft at high velocity, could not conceivably add enough energy to a satellite to heat it appreciably—it would be like trying to heat up a battleship by firing one small but very hot bullet at it each second.

Actually, the temperature of a satellite is determined primarily by radiant heat transfer and by conduction of heat between its various parts. When the satellite is in the sunlight it intercepts a flux of radiant energy amounting to about 1400 W/m^2†

† W is the abbreviation for the power unit Watt.

over a wide spectral range but with maximum energy in the visible region. Part of this energy is reflected, of course, and the remainder absorbed in the form of heat. The designer can control how much energy will be absorbed and how much reflected in the various parts of the spectrum by specifying different surface treatments or coatings for the satellite, ranging from dull black paint to highly polished mirror surfaces. On average, under steady-state conditions, the satellite must dispose of this heat energy plus any additional heat that may be generated on board by radiating it as infrared energy into empty space.

Radiation by the satellite

The radiant temperature of the clear black sky, looking away from the sun, earth, and moon is very low—approximately 3 K (−270 °C). Consequently any surface of the satellite which is shaded from these radiant sources can radiate very efficiently and thus become quite cold unless heat is supplied by conduction or convection from some warmer part of the satellite. The temperature of the earth, which subtends a large solid angle at satellite altitudes, varies between about 288 K (+15 °C) for the surface and 218 K (−55 °C) for the highest clouds. Satellite temperatures normally fall somewhere in this range, and so a satellite surface facing the earth receives about as much radiant energy from the earth as it radiates back to it, thus the net transfer from such a surface is small. The moon is so small as seen by an earth satellite that its effect on the gross thermal balance is negligible.

For black bodies—that is, objects that absorb all of the incident radiation at all wavelengths—the total radiated energy varies as the fourth power of the temperature. The spectral distribution of this radiation is characterized by an increase in radiated energy with decreasing wavelength until a peak is reached, after which the radiated

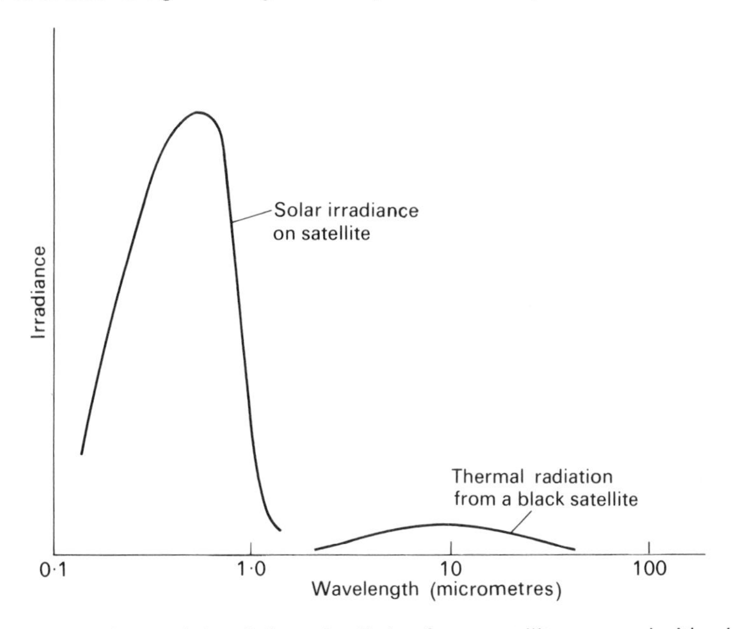

Fig. 3.3. Spectral characteristics of thermal radiation from a satellite compared with solar irradiance on a satellite.

energy falls off again. As the temperature of the body is raised the radiated energy increases at all wavelengths but more rapidly at the shorter wavelengths so that the peak shifts toward the short wavelength end of the spectrum, as, for example, from infrared to visible wavelengths. The wavelength of the maximum radiation is inversely proportional to the absolute (Kelvin) temperature.

Of course, no object is a perfect black body. Every object with which we are familiar reflects some energy. (In the laboratory a fairly good approximation to the black body is created by using a small open hole in the side of an otherwise closed box with rough interior walls.) The ratio of the energy actually radiated at any wavelength to that which would be radiated by a hypothetical black body at the same temperature is called the emissivity, and is always less than 1·0. In general, the emissivity for any given surface is the same as the absorptivity, which is 1·0 minus the reflectivity. The emissivity of most materials varies considerably with wavelength. For example, certain white paints show very low emissivity (high reflectivity) in the visible part of the spectrum, but high emissivity (absorptivity) in the infrared.

Fig. 3.3 shows in a relative way the different spectral characteristics of the solar irradiance falling on the satellite and the thermal energy which it radiates. The former peaks at about 0·5 micrometres in the visible part of the spectrum; the latter at about 10 micrometres in the infrared. The total radiant energy for a black body at 300 K (a high but not unreasonable satellite temperature) can be calculated and is found to be about 460 W/m^2 compared with the solar irradiance of about 1400 W/m^2. Thus if its emissivity were the same for both visible and infrared radiation, a simple spinning satellite uniformly exposed to one-half solar radiation (on the average) and radiating continuously to space over all its surface would be unable to maintain a temperature as low as 300 K and would tend to become hotter. This was indeed a problem in the design of early satellites; it was solved by using appropriate paints or mechanical surface treatments to increase the reflectivity at visible wavelengths while simultaneously increasing the emissivity at infrared wavelengths.

Hot spots

After gross thermal balance has been achieved, the designer must still be concerned about local hot spots where energy is dissipated by electrical equipment inside the satellite. In unpressurized satellites such problems are usually solved by providing good thermal conductivity between the point of heat dissipation and an emitting surface (radiator) on the outside of the satellite in an area that is shielded from the sun. Heat pipes are also sometimes used to transfer thermal energy by convection from the point of dissipation to the radiating surface. Some measure of temperature control can be achieved by using movable shutters to change the characteristics of the radiating surfaces. All of this rapidly becomes very complicated and usually requires the services of thermal-design specialists and the use of sophisticated mathematical models which can be run on electronic computers each time even a small change is made in the spacecraft configuration.

Re-entry vehicles

When it is necessary for a satellite or a part of a satellite to re-enter the earth's atmosphere without damage to its contents (as in the case of a manned satellite) it is called a re-entry vehicle. In this case atmospheric heating becomes quite large and is indeed the most critical problem for the designer. It can be overcome by covering the forward part of the re-entry vehicle with a specially designed heat shield made of a material like graphite that can withstand a very high temperature for a short time. This material is backed by a good thermal insulating material to keep the heat from penetrating inside the vehicle. Fortunately the high temperature does not last very long because as the atmospheric heating builds up so does atmospheric drag. Thus the vehicle slows down quickly and in a few minutes the heating is reduced to a low value.

Problems of electromagnetic radiation

In addition to the visible and infrared emissions from the sun and earth, the thermal effects of which have already been considered, radio waves, ultraviolet radiation, and X-rays are also present in the satellite environment and must be considered by the designer of the satellite and its communication subsystem.

At wavelengths of less than a metre, which are normally used for communication between the satellite and earth stations, the radio noise of the atmosphere and the Galaxy is of little consequence, except possibly to the scientist who may be trying to measure it as part of a scientific experiment. Even the sun, which is a relatively powerful radio emitter, is not a significant source of interference in the design of most satellite communication systems, although it can become significant for deep-space systems where communication must be maintained over distances on the order of a hundred million kilometres or more. To put this situation into proper perspective—when the sun is relatively undisturbed the power flux in the vicinity of the earth is about 10^{-20} W/m^2 per hertz for wavelengths in the vicinity of 0·25 m. Bursts of noise associated with intense solar disturbances can reach 10–100 times this level. However, a 10-watt transmitter with an antenna beam width of about 6°, modulated in such a way as to spread its energy uniformly over a bandwidth of 10 kilohertz would create a flux on the order of 10^{-13} W/m^2 per hertz at a range of 1000 kilometres. This is ten million times greater than the power flux of the quiet sun.

Solar ultraviolet and X-ray emissions normally do not penetrate the metal skin of a satellite. Only during solar flares are X-rays observed having sufficient 'hardness' to penetrate a few millimetres of aluminium, and the intensity and duration of these occurrences is not sufficient to damage equipment inside the satellite. Some polymers (paints and plastics) and optical elements could be damaged by exposure to this radiation and therefore should not be used on the outside of a satellite. Such materials are even more likely to be damaged by energetic charged-particle radiation, which will be discussed later in this chapter.

The photoelectric effect

The principal concern of the satellite designer, in so far as ultraviolet and X-rays are concerned, is with the photoelectric effect which they produce on the sunlit outer surface of the satellite. Almost all materials will emit photoelectrons when irradiated by photons in this energy range (approximately 10 electron volts to 10 kiloelectron volts—see p. 140). Consequently, because of the emission of negatively charged electrons, a sunlit spacecraft will tend to acquire an electric potential somewhat more positive than that of the ionized atmosphere through which it moves. The charging process will continue until the spacecraft will have surrounded itself with a dense enough sheath of electrons to repel the lower-energy photoelectrons and cause them to return to the spacecraft surface. This phenomenon is analogous to the space-charge effect in thermionic devices. The electric potential of the spacecraft is thus limited by the space-charge effect and by the conductivity of the surrounding ionosphere to a few volts or, at most, a few tens of volts. Such a potential difference is of no practical consequence for normal spacecraft systems, but is of considerable concern when instruments are carried on board for the measurement of the low-energy electrons in the solar wind because the motion of such electrons can be greatly modified by small potential gradients.

The photoelectron 'sheath' which surrounds the spacecraft will also change the impedance of any radio antenna located on the spacecraft and must therefore be taken into account in the antenna design.

The geomagnetic field

As everyone knows, the earth is a magnet. It has north and south magnetic poles and an external field that is approximately that of a short, stubby magnetic dipole, slightly tilted and offset with respect to the earth's rotational axis, as shown in Fig. 3.4. When magnets first became the subject of scientific curiosity in the latter part of the sixteenth century, the two poles of a bar magnet were identified as 'north-*seeking*' and 'south-*seeking*' poles because of the positions they assumed when the magnet was allowed to turn freely in the earth's magnetic field. These designations later became simply north and south poles, which is somewhat confusing

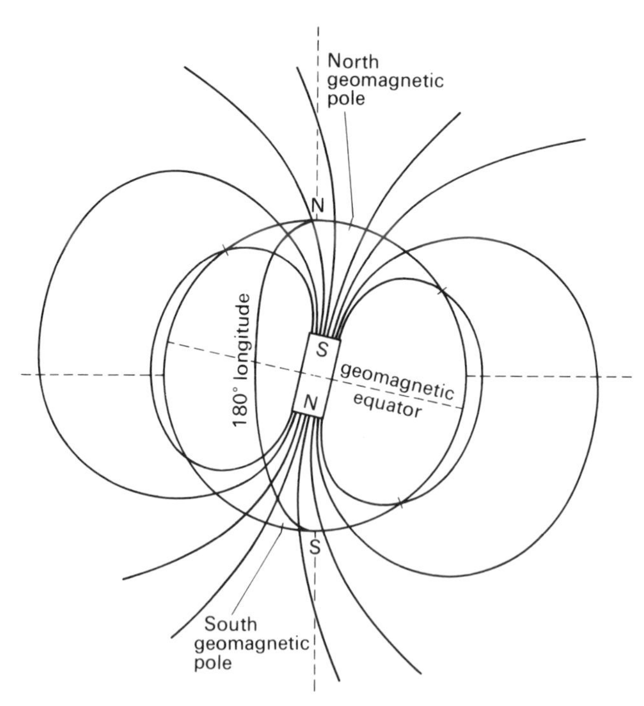

Fig. 3.4. Dipole model of the earth's magnetic field. View is downward in the middle of the Pacific Ocean. The equivalent dipole is shifted about 436 kilometres from the centre toward the observer and tilted about 11·5°, as shown.

because the points on the earth's surface normally referred to as the north and south geomagnetic poles are, by this definition, actually south and north magnetic poles, respectively. We must simply learn to live with this confusion! It is generally believed that the main geomagnetic field is produced by a phenomenon similar to that in a self-excited dynamo, involving thermally driven convection in an electrically conducting molten core deep inside the rotating earth. The eccentric dipole field is modified by various anomalies, some of rather large extent, thought to be associated with irregularities in the internal hydromagnetic dynamo system, and others of lesser extent, thought to be caused by irregularities in composition of the earth's crust. It is also subject to slow secular changes that make it necessary to modify periodically the magnetic charts used for navigation.

The magnetic field caused by phenomena inside the earth is modified by electric currents flowing in the ionosphere and by interaction with streams of plasma shot out from the sun. These transient disturbances, when unusually strong, are referred to as magnetic 'storms'. Although they may adversely affect navigation and communication circuits the short-term variations in the geomagnetic field amount to only a few tenths of a per cent at low altitudes. At altitudes on the order of 10 000 kilometres or greater both the transient geomagnetic phenomena and the distortion of the main field by solar 'wind' pressure become relatively much more significant.

The satellite designer is concerned with geomagnetism for four reasons. First, an electrically conducting satellite spinning in the earth's magnetic field will generate induced electric currents that slowly decrease the spin rate. Also, even if it is not spinning, the motion of the satellite through the magnetic field creates an electromotive force so that opposite extremities of a large satellite have an electrical potential difference on the order of one or two volts. If the satellite is moving through the ionosphere, which is conductive, an electric current driven by this potential difference flows through the satellite and out into the ionosphere. The interaction of the current with the magnetic field generates a force opposed to the motion that is normally very small but has been found to be significant in large, light-weight satellites such as the *Echo* metallized balloon satellites launched by the United States during the early 1960s. Second, the designer may wish to use the earth's magnetic field either as an earth-fixed directional reference system, or as a means of producing torque for controlling the attitude of the satellite. Third, if the satellite carries sensitive magnetic instruments for measuring the magnetic field in its vicinity, the designer must be careful to keep all stray magnetic fields generated by electrical and magnetic devices and subsystems within the satellite small as compared with the natural field which it is desired to observe. And fourth, the geomagnetic field determines the environment of energetic charged particles in the vicinity of the earth, which can affect satellite operation in several important ways.

The charged particle environment

The most intense penetrating radiation now known to be present in the vicinity of the earth is to be found within the trapped radiation belts, or Van Allen belts as they are frequently called. Early in 1958, Professor James Van Allen first identified geomagnetically trapped corpuscular radiation in his observations with a shielded Geiger counter on the American-launched *Explorer 1* satellite, although trapping of charged particles had previously been predicted on theoretical grounds by Alfvén and others.

The Van Allen belts

The Van Allen belts, which lie just outside the earth's atmosphere, contain protons with energies up to several hundred megaelectronvolts and electrons with energies up to perhaps 5 megaelectronvolts. The motion of these charged particles is constrained by the geomagnetic field and is rather complex. In a very simplified analysis it can be divided into three components. First, there is a circular component in a plane perpendicular to the direction of the magnetic field. The radius of this circular motion depends on the energy, mass, and charge of the particle and on the field strength, but is always much less than the radius of the earth. Second, if the particle velocity at the magnetic equator, where the field lines are essentially parallel, has a component in the direction of the field lines, the centre of the circular motion (sometimes called the guiding centre) will move down along the field lines until the field strength becomes large enough to prevent further penetration. It will then reverse and move back along the field lines to a point at the same field strength in the opposite hemisphere. Thus the particle continues to bounce back and forth along a spiral path between 'mirror' points in the northern and southern hemispheres. The position of the mirror points depends on the initial pitch angle of this spiral. If it lies too deep within the outer atmosphere, the particle has a high probability of being removed by collision. Third, as the particle bounces back and forth, its path drifts slowly in longitude—protons to the westward and electrons to the eastward. If the geomagnetic field were a perfect dipole, the guiding centre would thus trace out a shell which would be the figure of revolution of part of a single field line. In the actual geomagnetic field, the particles tend to descend to lower altitudes over negative geomagnetic anomalies, the largest of which is located in the southern Atlantic.

In order to describe the charged particle environment, it is necessary to give the intensity in terms of the number of particles which pass through an imaginary surface in a given time (omnidirectional intensity simply means that particles coming from all directions are included) and the energy or energy range of the individual particles. Particle energy is measured in electronvolts (eV), thousands of electronvolts (kiloelectronvolts, keV), millions of electronvolts (megaelectronvolts, MeV), or thousands of millions of electronvolts (gigaelectronvolts, GeV). An electronvolt is the energy a single electron would gain if it were accelerated through a potential difference of one volt.

The omnidirectional intensity of the hard proton component of the Van Allen radiation (40–110 MeV) has a peak of about 4×10^4 protons per square centimetre per second, in an equatorial ring at about 3200 kilometres altitude. It is about 100 times less at 1000 kilometres over the equator or at 600–700 kilometres over the South Atlantic, and is essentially negligible at any altitude at latitudes greater than 50° north or south. The electron component is much more variable and its distribution more complex. It overlaps the proton component, but usually tends to have a maximum intensity of around 10^8–10^9 electrons per square centimetre per second for energies greater than 20 keV at 10 000–15 000 kilometres altitude and falls off less rapidly with latitude than the proton component. It is, nevertheless, essentially negligible beyond 60° north or south.

Man-made disturbances

On 9 July 1962, a significant perturbation in the trapped electron distribution was introduced by the high-altitude atomic test, conducted by the United States, called

Starfish. The maximum intensity of this artificial increment was in the region extending from about 1600 kilometres to 6300 kilometres over the equator. At low altitudes this increase decayed rather rapidly, but at altitudes above 1800 kilometres the increase was still something like a factor of 10 several years after the test.

Solar flares

Next in importance to the Van Allen trapped particles, are the protons and electrons ejected from the sun during solar flares. About half an hour or more after the occurrence of a large flare has been detected by optical or radio instruments on earth, protons having energies as high as 200 MeV are likely to begin to arrive at the earth, primarily in the regions close to the magnetic poles. Protons in this energy range are stopped by collisions in the atmosphere and do not reach the surface; however, the peak flux above the atmosphere can be as high as $1–2 \times 10^4$ protons per square centimetre per second for energies greater than 30 MeV, and 5×10^3 protons per square centimetre per second for energies greater than 100 MeV. Although these bursts of protons usually last only a day or so, the integrated intensity over a single burst can be as much as 10^9 protons per square centimetre for energies over 30 MeV, and 10^8 protons per square centimetre for energies more than 100 MeV. Proton events of this kind occur at a frequency of 5 or 10 per year during the maximum activity years of the solar cycle, often in groups of 2–4 in fairly rapid succession.

Occasionally a solar flare will produce protons with energies greater than a gigaelectronvolt. These are called relativistic solar-flare events, because the velocity of the high-energy protons ejected is so close to the velocity of light that the theory of relativity must be used to compute their motion correctly. Particles of this energy can sometimes penetrate the atmosphere far enough to be detected on the ground and they also are deflected to a lesser extent by the geomagnetic field and are therefore less concentrated at the magnetic poles, but they generally do not reach the earth at low latitudes.

Cosmic rays

Protons and a few heavier particles having energies in the range from about 0·1 GeV to thousands of GeV come to the earth from outside the solar system and perhaps even from outside our Galaxy. These are known as cosmic rays. Like the higher-energy solar protons, which are often called 'solar cosmic rays', the galactic cosmic rays of lower energy, relatively speaking, are deflected to some extent by the geomagnetic field toward the polar regions. Higher-energy cosmic rays, however, can penetrate to progressively lower latitudes and the most energetic of these can be detected at terrestrial observatories near the equator although such events are relatively infrequent. The directional intensity of cosmic rays having energies in excess of 0·1 GeV is normally about 0·2 particle per square centimetre per second per steradian[†] at times of high solar activity and about 2·5 times higher at times of minimum solar activity. During one of the relativistic solar-proton events mentioned above, this level can be temporarily increased by a factor of several hundred to several thousand by the additional flux of high-energy particles from the sun, which behave in much the same way as galactic cosmic rays.

It should perhaps be mentioned here that rather intense flows of electrons and protons having energies on the order of 100 keV are sometimes observed at very high

[†] A steradian is a measure of solid (or three-dimensional) angle.

altitudes over the equator and at lower altitudes over the polar regions. These particles do not appear to be trapped but rather to reach the earth along magnetic field lines that have somehow been 'broken' and reconnected into the interplanetary magnetic field. They are generally observed during periods known as magnetic substorms. When they descend into the upper atmosphere they cause the beautiful auroral displays that are often seen at such times. Satellites in geosynchronous or stationary orbits (at an altitude of about 36 000 kilometres) sometimes encounter clouds of these particles, especially when near the morning meridian, and have been damaged by electrical discharges resulting from the charge they build up on satellite surfaces. In lower orbits the electric charges tend to be drained off by the conductivity of the atmosphere.

Shielding the satellite

The most important aspect of the energetic particles encountered by a satellite is their ability to penetrate solid materials and to cause temporary malfunctions in electronic circuits or permanent damage to these and other parts of the satellite system. When a charged particle interacts with a material such as the aluminium skin of a spacecraft, the principal effect is to create disturbances in the energy levels of the atomic electrons in that material, frequently knocking one or more electrons completely out of any atom with which it interacts. This process is referred to as ionization. If the momentum of the particle is great enough, it may even dislodge complete atoms from their positions in the solids. At very high (relativistic) energies

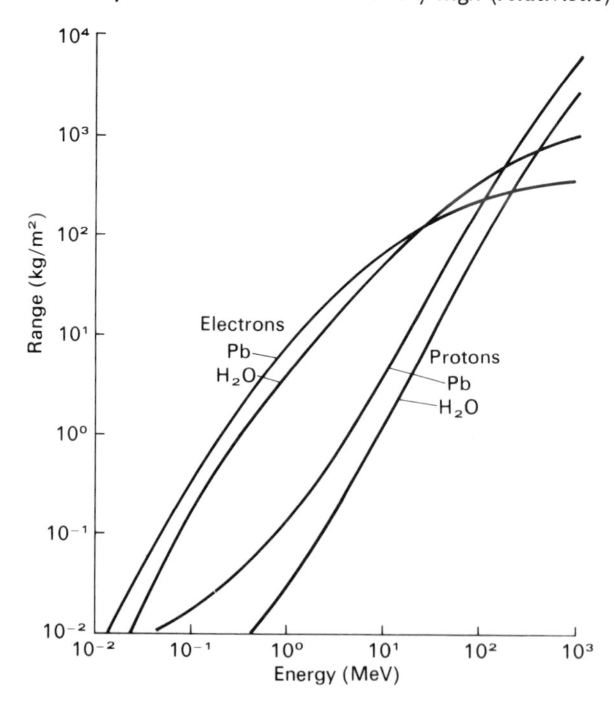

Fig. 3.5. Mean range of electrons and protons in lead and water as a function of energy.

nuclear interactions can take place with the production of neutrons and other secondary particles. When the velocity of an electron is suddenly changed, it radiates a fraction of its energy in the form of electromagnetic waves through a process known as *Bremsstrahlung*, the fraction being proportional to the electron energy and to the atomic number of the material with which it interacts. All of these interactions, of course, remove energy from the incoming particle which, if the material is thin, emerges at lower energy than it entered. Otherwise it eventually comes to rest within the material.

The distance a particle can travel in a material multiplied by the density of the material is referred to as the 'range' of the particle in that material. The relationship between mean range and particle energy for electrons and protons in lead and in water is shown in Fig. 3.5. For light-weight engineering materials such as aluminium, magnesium, or plastics, the data will fall between these limits, generally closer to the curve for water.

The range of a 1 MeV electron can be seen to be about 0·5 g/cm^2. Thus a thickness of about 2 mm of aluminium will stop most of the electrons to be found in the Van Allen belts. Shielding of more than this effectiveness is normally provided by the mechanical structure of a satellite. Approximately 1 per cent of the energy of the stopped electrons will, however, be transformed into soft X-rays by the *Bremsstrahlung* effect. These X-rays are more penetrating in low-atomic-number materials such as aluminium than are electrons, thus additional shielding—preferably of a high-atomic-number material such as lead—may be required to attenuate them. Protons of energy 10–100 MeV have a mean range of about 0·17 g/cm^2 to about 10 g/cm^2 respectively (0·06 to 3·7 cm of aluminium). The larger ranges begin to pose a very difficult problem in terms of shielding weight. Thus the higher-energy protons trapped in the geomagnetic field or ejected from the sun during large solar flares emerge as the main concern of the satellite designer, with *Bremsstrahlung* X-rays from trapped electrons as a secondary but sometimes still significant consideration. Shielding against cosmic rays is, of course, out of the question, but fortunately the intensity of this type of radiation is low enough not to be troublesome except for very long-life satellites at great distances from the earth.

Radiation damage to organic materials occurs primarily by the ionization process through chemical reactions or alteration of molecular structure, the nature of which depends on the material involved. Thus a plastic like Teflon which scissions (breaks chains) easily will decrease in strength when exposed to ionizing radiation, whereas polyethylene, which tends to cross-link rather than to scission, will increase in strength and toughness, at least up to a point. Crystalline materials such as semiconductors are damaged predominantly by atomic displacements; however, semiconductor devices such as transistors can be affected by ionizing radiation through the deposition of electric charges on surfaces or within insulating layers where they are trapped for considerable periods of time. Such an effect caused trouble in the early *Telstar* communication satellite before it was well understood.

Micrometeoroids—cosmic dust

Meteoroids are particles of stony or metallic material ranging in size from microscopic specks to large chunks. The probability of encountering such a particle decreases with size, as shown in Fig. 3.6. Thus there would be only an even-odds chance that a satellite having a cross-sectional area of a square metre would be struck

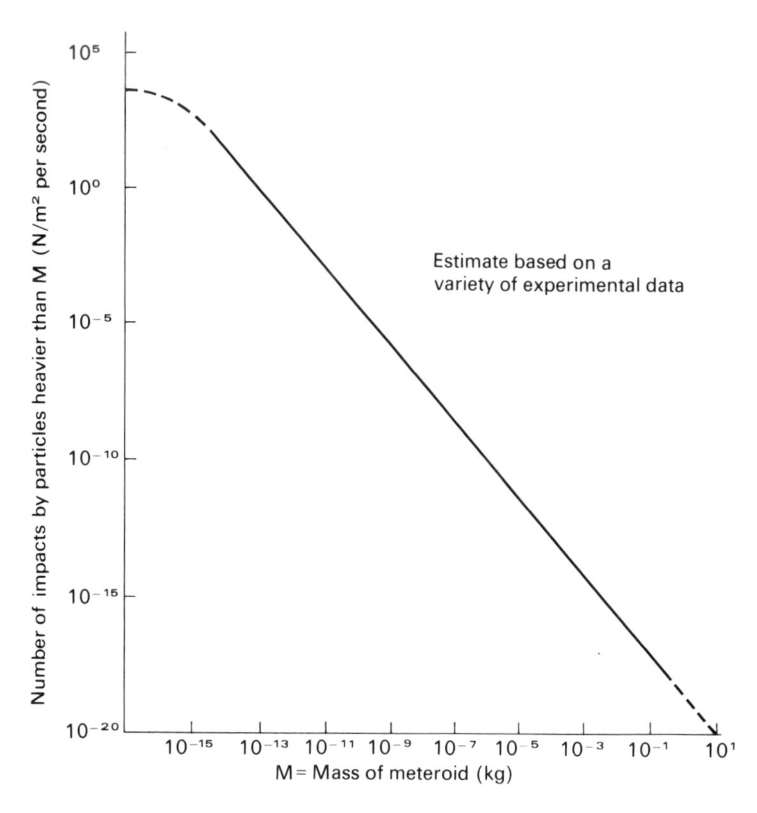

Fig. 3.6. Cumulative mass distribution of near-earth meteoroid flux.

by a particle more massive than a tenth of a milligram (about 0·3 mm to about 0·6 mm in diameter, depending on density) in 10 years (approximately 3×10^8 s) of orbiting. A particle of this size, upon entering the atmosphere, disintegrates or burns up, leaving behind an ionized trail that is just observable by radar and photographic methods. The bright meteors or 'shooting stars' which we can see with the naked eye on a clear night are at least 1000 times more massive and 10 times bigger in diameter.

Still larger chunks of meteoric material occasionally reach the surface of the earth. These are believed to be fragments of asteroids and are generally divided into two main categories: the stony meteorites, and the much-less-frequent iron meteorites. Their chemical composition and physical properties are well known. However, it is believed that the smaller particles, which are of interest to the satellite designer, are mostly of a different type and have their origin as cometary debris. It is extremely difficult to study these particles because, like the eligible bachelor, they are so hard to catch. However, they seem to be rather loosely bonded aggregates of iron, iron oxide, nickel, and metallic silicate grains, having a density in the range 0·1–0·5 g/cm³. Their velocity with respect to the earth ranges from about 11 km/s to 72 km/s. Some, of course, may be trapped in earth orbit, and these could have a lower velocity with respect to a satellite; however, an encounter velocity of about 30 km/s is considered to be typical.

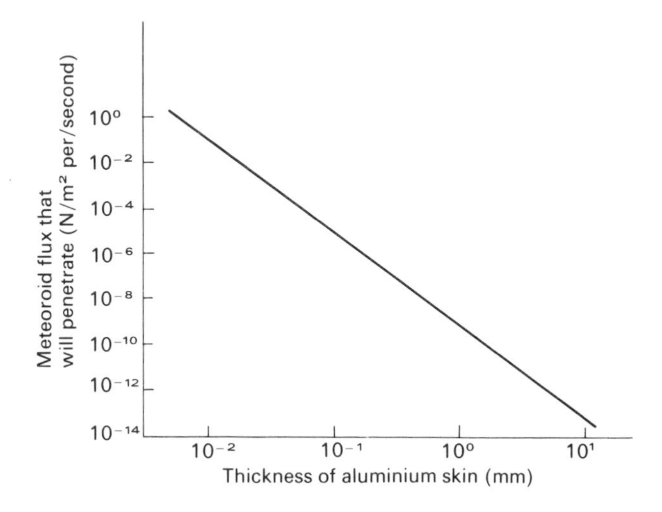

Fig. 3.7. Estimated meteoroid penetration of thin aluminium in near-earth space.

At such speeds even a snowflake would be a dangerous projectile. Unfortunately there is no rigorous theoretical basis on which to predict the damage that will result from a meteoroid impact. Engineers must 'make do' with estimates or empirical formulae which represent reasonable extrapolations of a relatively few laboratory tests at velocities in the range of 10–20 km/s. Fig. 3.7 shows the expected mean flux of particles believed to be capable of penetrating various thicknesses of aluminium sheet. For steel sheet, the penetrations would be less by about an order of magnitude.

It must be realized that the engineer cannot design his satellite on the basis of average or mean values of meteoroid penetration frequency any more than an architect can design a hotel or office building with an elevator capacity only sufficient to handle the average number of passengers per hour. He must therefore obtain or assume a distribution of the data about the mean, and on this basis attempt to determine the thickness of shielding required to provide a desired level of reliable protection against meteoroid penetration. In the end, he must accept some probability that a meteoroid will penetrate and damage some vital part of the satellite. Fortunately, especially in the case of manned satellites, this probability can be very small.

Cost constraint

Clearly the satellite environment is not only an unfamiliar one to earthly engineers but also, in many respects, a downright unfriendly one. Nevertheless, there does not seem to be any aspect of this environment for which satellites could not be designed to have any reasonable level of reliability and life, provided there were no limit on weight or cost. However, such limits do exist. The satellite designer must make exacting compromises to achieve the greatest possibility of success for the desired mission within the imposed constraints. To the extent he accomplishes this he deserves indeed to be praised.

4 An aid to communication

. . . since good, the more communicated more abundant growes.
John Milton (1667) *Paradise lost,* Book V

Presumably the angel in Eve's dream in the above quotation could have added that the same statement might also be made about evil, and that it seems especially applicable to nonsense and trivia. The ability to communicate is clearly neither good nor bad, except in as much as it makes more available or more 'abundant' that which is communicated. Computer experts have a phrase which sums it up rather neatly in only four words! 'Garbage in, garbage out!'

The distinguishing feature of 'intelligent' man seems to be his ability to create languages or codes by which complex and abstract ideas can be communicated from one individual to another or to many others and can be stored so as to be communicable even to future generations. In the beginning, such communication must surely have been vocal in nature. Perhaps this is the reason for the dual meaning of the word 'spirit' in many languages—it means wind or breath, but also life in the intellectual sense. In any case, *Homo sapiens* is a 'windy', talkative creature, and one of his most avid occupations since the beginning of time has been to increase the speed, range, and diversity of his ability to communicate.

A brief history of communications

After the development of spoken language the next great step was writing, which made it possible to transmit messages beyond the vocal range of the sender and to record information for long periods of time. Drums, smoke signals, flags, semaphore, and other means were used to speed the delivery of messages, especially through the jungle or over water where messengers could not travel very swiftly. Printing, reputedly invented by the Chinese in A.D. 770, became a practicable process in the Western world about 600 years ago through the work of Johannes Guttenberg, and made mass communication possible for the first time. The development of electrical technology led to the invention of the telegraph, one form of which was patented in England by Sir Charles Wheatstone in 1837 and another in the United States by Samuel Morse in 1840. The latter system, as finally developed, used a remarkably ingeneous code (the Morse code) consisting of short and long pulses of electric current combined in such a way as to represent the most commonly used letters of the alphabet by the shortest combinations. Printing and the telegraph were combined when a Washington newspaper began to print 'Telegraphic News' in 1844. The telephone was not far behind. Following Alexander Graham Bell's first demonstrations in 1876, commercial lines and exchanges were opened in 1877 and 1878.

Meanwhile, communication by pictures had developed in parallel with communication by language. It had progressed from early Paleolithic cave drawings, through the brilliant art forms of developing civilizations, to the development of photography in 1839 by a French painter named Daguerre. The technique of photography, based on developments in optics and chemistry, made it possible to reproduce exact images of a face or scene and to make as many reproductions as desired.

Rapid communication across the oceans by means of submarine telegraph cable began in 1855, when Cyrus Field successfully laid a telegraph cable between Canada and Newfoundland. In 1858, he successfully extended his cable link to Ireland, thus connecting for the first time the continents of Europe and North America. Unfortunately this first cable failed after a few weeks and was not successfully replaced until 1865. Early telegraphic cables were relatively costly and exceedingly limited in capacity. It is reported that it required more than an hour to send from England to the United States Queen Victoria's cabled greeting to President Buchannan. Geomagnetic disturbances frequently put the cable system out of operation for hours or even days at a time. However, at that time it required about 20 days to send a letter across the Atlantic by ship, which was the fastest alternative mode of communication.

It was not until after the development of broad-band amplifiers reliable enough to be spliced into the cables at intervals and left to operate unattended for years on the ocean bottom that high-speed telegraphy and telephone signals could be sent through these intercontinental cables. Thus the first transatlantic telephone cable system did not go into service until 1956, almost 90 years after the first telegraph cable system!

Meanwhile a new technology called 'radio' had already made it possible to send both telegraphic messages and the sound of a man's voice across the oceans. In 1895 Guglielmo Marconi demonstrated the first wireless telegraph, and in 1900 Regenald Fessenden demonstrated wireless voice transmission. By December 1901, Marconi had demonstrated the transmission of radio telegraph signals across the Atlantic. Intercontinental radio telegraph and radio telephone systems were operating routinely by 1920 and radio broadcasting, a major new industry, had been born. The sending of pictures by wire and radio (facsimile transmission) came into widespread use in the late 1920s and early 1930s, as did the teletypewriter, which reproduced typewritten messages much faster and more accurately than human telegraph operators using Morse code could do. Also in the 1920s embryonic experiments were being made with something called television—the broadcast transmission in real time of moving images from a special electronic camera. By the late 1930s this technique had developed to the point where the pictures had an acceptable quality, and following the Second World War it became a commercial service, with profound social, cultural, political, and economic consequences. Also, as a direct result of radar developments during the Second World War it became possible to use radio at centimetre wavelengths (called microwave radio), in the form of beams only a few degrees wide, to communicate from point to point with greatly increased information-handling capacity, efficiency, and reliability.

With the development of high-speed digital electronic computers during the 1950s and 1960s there has been a growing demand for high-speed communication of data in digital form. Fortunately, the technologies developed by communication engineers have been adequate to meet this need, so that data transmission can now be achieved at speeds up to hundreds of thousands or even millions of bits (binary digits) per second. Thus a new service has been born that is concerned with enabling computers to 'talk' to each other in their own esoteric languages at speeds almost beyond the capability of man to comprehend. A new technology based on the use of infrared or light waves, generated by lasers and transmitted over optical fibres or in very narrow coherent beams through space, is currently being developed to extend this capability for high-speed communication even further.

The latest development which is being used to help satisfy the rapacious human craving to communicate is space technology.

We might well ask why any further developments were needed. Transoceanic telephone and telegraph communication had already been achieved by radio and cable; television programmes had been sent thousands of miles across continents by microwave relay or coaxial-cable techniques. What more was required? Well, for one thing, there was no really satisfactory transoceanic television capability. Although attempts to relay television signals across the Atlantic by cable and radio had been partially successful, even black-and-white pictures were of poor quality and transatlantic colour-television transmission seemed to be almost out of the question. Also, long-distance radio circuits depended on reflection by the ionosphere, which varied in height, intensity, and structure with time of day, season, and solar activity. Voice transmission was often distorted and made to sound funny by selective fading. Sometimes, following a solar disturbance, these circuits would go out completely for hours or even days. The part of the radio spectrum that can be used for this 'sky-wave' mode of transmission extends from about 5 megahertz to about 15 megahertz at night and to 25 or 30 megahertz in the daytime. Even if all these frequencies could be used exclusively for radio telephone all the time, which is clearly not possible, there would be room for only a few thousand circuits worldwide. Submarine cables provided service of better quality and reliability; however, the cost was high and the capacity still inadequate to handle the available traffic.

Thus the communications industry could only look to the higher-frequency part of the spectrum—above 50 megahertz—for room to grow. Radio waves at these frequencies are not significantly reflected by the ionosphere; they pass on through it into space. They will not, however, pass through the solid earth. Hence transmissions in this region must be essentially by line of sight: that is, it must be almost possible to see the transmitting antenna from the receiving antenna except for such tenuous obstacles as trees. Because the earth is round, this line-of-sight requirement limits transmissions to a rather short distance, unless the antennas at both ends are at high elevations. That is one reason why the operators of television transmitters like to put their antennas on top of the tallest buildings or towers available.

When it began to seem technologically possible to establish artificial satellites in orbit around the earth communication experts were quick to realize that they could be used as 'super-towers'—platforms for radio-relay stations so high that they could provide line-of-sight communications between points a third of the way around the world. Because they could use centimetre wavelengths and highly directional antennas they would require relatively low radiated power, despite the great distances involved. At these frequencies plenty of bandwidth would be available to relay many television signals and tens of thousands of telephone conversations simultaneously. It is perhaps not surprising therefore that communication by satellite was the first utilitarian purpose to which the new space technology was applied.

The development of satellite communication systems

Passive and active systems

The earliest attempts to communicate via satellites involved 'passive' systems—that is, systems using simple radio reflectors in orbit to reflect or scatter microwave beams from one earth station back to another. Two large metallized balloons, *Echo 1* and

Echo 2, were orbited for this purpose in mid 1960 and early 1964 respectively. The second balloon was part of a co-operative programme and was used to transmit signals between the Jodrell Bank Observatory in England and the Zimenki Observatory in the Soviet Union. In 1963, a belt of orbiting dipoles (hair-thin wires a few centimetres in length) was created and used for military communication experiments. This latter experiment, known as West Ford, created something of an uproar among radio astronomers, who anticipated that if it were successful the sky would be filled with these scattering elements, making astronomy nearly impossible. The advantage from the military viewpoint was that such a system could provide a communication link that would be almost impossible to jam. The experiment was quite successful but, partly out of deference to the astronomers, it was not repeated.

Concurrent development of active satellite relay systems proved so successful that all further work on passive systems was eventually dropped.

Live by satellite

The caption 'live by satellite' has now become so commonplace on television screens in our homes that it hardly seems necessary any more. It was first used after the *Telstar 1* satellite was launched in July 1962. This satellite, which was designed and constructed at the Bell Telephone Laboratories and launched by NASA, provided an opportunity for hundreds of communication experiments and demonstrations, including the first high-quality live television transmissions across the Atlantic. It failed at one point because of the deleterious effect of trapped radiation on its transistors, but was partially 'revived', and was replaced in May 1963 by the more reliable *Telstar 2.* By this time wide-band transmissions were being made on a fairly frequent basis between a ground station at Andover, Maine and stations at Goonhilly Downs, England and in Plemeur-Bodou, France.

By December 1962, the NASA *Relay 1* satellite, constructed by RCA, had joined *Telstar* in orbit and was also being used for communication experiments and demonstrations. *Relay 2,* designed for longer life and higher reliability, was added in January 1964.

Geostationary satellites

Both the *Telstar* and *Relay* satellites were in relatively low orbits and, although the communication quality was good, each satellite was within line of sight of any pair of ground stations for only a few hours each day and it was necessary for the highly directional ground-station antennas to track the satellites as they moved rapidly across the sky. In Chapter 2, the possibility of a geostationary satellite orbit was mentioned. Such a satellite could remain almost stationary in the sky, always in view of the ground stations at both ends of the communication link and requiring little or no tracking capability. NASA began to experiment with earth-synchronous communication satellites in 1963. The first attempt, in February, failed. The second, in July 1963, successfully placed *Syncom 2* in an earth-synchronous orbit, having a period of exactly 24 hours but with an inclination of about 38°, so that it traced out a figure-of-eight pattern in the sky, ranging from 38° N in latitude to 38° S in latitude each day. This tiny satellite weighed only about 40 kilograms, including a small propulsion system for keeping its spin axis aligned and causing its orbit to drift to the desired location. Its transmitter used only a two-watt travelling-wave tube at 1815

Table 4.1. Progress in commercial satellite development

	Intelsat I (*Early Bird*)	*Intelsat II*	*Intelsat III*	*Intelsat IV*	*Intelsat VIa*	*Intelsat V*
Year of first launch	1965	1967	1968	1971	1975	1978–80
Satellite weight in orbit (kilograms)	39	87	150	722	730	900
Launch vehicle	Thor–Delta	Improved Thor–Delta	Long-tank Thor–Delta	Atlas–Centaur	Atlas–Centaur	Atlas–Centaur
Average number of circuits†	240	240	1500	4000	5550–6000	To be determined
Design life-time (years)	1·5	3	5	7	7	7
Coverage	Hemisphere	Near global	Global	Global two-spots	Global plus two hemispheres	
Main technical advancements	Commercial service initiated	Multiple access	World-wide service initiated	Major path concept	Shaped new beams	New frequencies TDMA‡

† Capacity may be greater or less depending on technical and operating factors.
‡ Time-domain multiple access.

megahertz, but it was able to relay voice, facsimile, and teletype signals between stations on both coasts of the United States and a ship as far east as Lebanon. Television signals were successfully transmitted from Fort Dix, New Jersey, to Andover, Maine.

In August of 1964,·NASA put up the first true geostationary communication satellite, *Syncom 3,* located at longitude 180° W over the equatorial Pacific Ocean. All of the commercial satellites, which began to be launched in 1965, have been geostationary satellites.

Commercial satellites

Communication by satellites began to acquire commercial status in the United States with the passage by Congress of the Communications Satellite Act of 1962 and the creation of the Communications Satellite Corporation (*Comsat* Corp.) in 1963. In 1964 it was made a subject of international commerce by the establishment of the International Telecommunications Satellite Consortium (*Intelsat*) with an initial membership of 14 national telecommunication authorities. At the time this is written, only some 10 years later, *Intelsat* has grown to a membership of 91 national authorities. Beginning with one satellite (*Early Bird*) and 240 voice circuits in 1965, the space segment of this global satellite system now has four operational *Intelsat IV* satellites, each with a potential average capacity of about 4000 telephone circuits, plus international colour television channels. There are also several older satellites, still useful to provide back-up capability. As of the end of 1974, regular full-time commercial satellite services were being provided world-wide by these satellites and by a network of 85 antennas at earth stations in 52 countries. The satellites have a multiple-access capability that makes it possible for them to operate simultaneously with several earth stations. There are now more than 293 recognized satellite communication pathways among countries with earth stations, and the number is increasing dramatically. This fantastic decade of growth is illustrated by Tables 4.1 and 4.2 for the space and ground segments respectively.

Table 4.2. Growth of communication-satellite ground system

Year	*Countries with antennas*	*Number of antennas*	*Satellite pathways*	*Leased equivalent half-circuits*	*Television half-channel hours*
1965	5	5	1	150	80
1966	6	8	1	172	152
1967	11	15	10	1049	450
1968	13	20	19	1525	1372
1969	24	41	82	2984	1826
1970	30	51	131	4388	2428
1971	39	63	181	5834	3562
1972	49	79	255	7527	6792
1973	52	85	293	9814	6817
1974	60	104	359	11 507†	7361

†Does not include service through leased satellite transponders.

Intelsat IV

The *Intelsat IV* satellite, which currently provides the main part of the *Intelsat* space capability, is a spin-stabilized spacecraft. It has a mass of about 1384 kilograms at launch and about 730 kilograms in the desired geostationary orbit. The 654 kilogram difference in mass is primarily that of the solid propellant burned in the third-stage (or apogee) motor that is mounted in the satellite. The photograph in Fig. 4.1 gives us an idea of what this satellite looks like.

Fig. 4.1. Photograph of *Intelsat IV f* during preparation for installation in launch vehicle.

Attitude stabilization is required in order to make it possible to point the directional high-gain antennas on the satellite in the desired directions. The kind of attitude stabilization with which most of us are familiar is called three-axis stabilization. It makes use of small instrument-type gyroscopes or other sensors, together with inertia wheels or gas jets, to achieve stabilization independently about each of the three orthogonal axes used to describe the vehicle attitude. In an aircraft automatic pilot, these three axes would be referred to as roll, pitch, and yaw. Sometimes these terms are also carried over by anology into satellite usage, pitch being the angle about an axis of the spacecraft which is perpendicular to the orbital plane, roll the angle about a spacecraft axis which usually points in the direction of motion, and yaw the angle about a spacecraft axis perpendicular to the other two.

In the *Intelsat IV*, however, a different method, known as spin stabilization, is employed. The whole satellite, or at least most of it, is caused to spin about its axis of symmetry at a rate of about 60 revolutions per minute, so that it acts like a big gyroscope. The spin axis must be designed to be the axis of maximum moment of inertia and a small amount of mechanical damping must be provided to eliminate nutation (wobbling). Under these conditions, the satellite's spin axis remains fixed in inertial space with high accuracy for very long periods of time. It can, however, be precessed slowly to a different direction by the application of pulsed gas jets, located and timed in such a way as to produce the average torque required for the desired angle of precession. In this way, the spin axis of a satellite like *Intelsat IV* is aligned perpendicular to its orbital plane, which in this case is as nearly as possible coincident with the equatorial plane of the earth. The high-gain antennas are mounted on a platform that is 'despun', in other words, rotated continuously about the axis of the satellite in a direction opposite to the spin and at a speed such that the antenna beams appear to be stationary with respect to the earth. The whole platform can thus be adjusted to move the beams in an east–west direction by small changes in the phase of the despin drive. The individual antennas can be moved with respect to the platform in such a way as to produce east–west and north–south displacement of each beam with respect to the others.

The antennas carried by *Intelsat IV* consist of two 127-cm parabolic reflectors for spot-beam transmissions and four conical horns, two for hemispheric transmission and two for reception. Transmissions are in the 3707–4193 megahertz region, and reception is in the 5932–6418 megahergz region. The two spot-beams have a width of 4·5° and a gain of about 28·1 decibels, equivalent to a factor of about 683 in power over an omnidirectional antenna. They can be steered in 0·1° increments. Their purpose, of course, is to provide more bandwidth and therefore more communication capability with less radio-frequency power and less interference for regions of limited size, such as Western Europe or the north-eastern United States, that are characterized by high communication traffic densities. The horn antennas have a beam width of 17° and a gain of 16·7 decibels, equivalent to a factor of about 47 over an omnidirectional antenna.

The electronics system comprises 12 separate microwave transponders, each having its own individual travelling-wave-tube amplifier with six watts output and a maximum frequency deviation of 36 megahertz. Thus the total bandwidth used by satellite is 432 megahertz. Maximum capacity in the spot-beam mode is 9000 two-way telephone circuits, or 12 colour-television transmissions, or some combination of both. For an average mix of services using both spot-beam and global coverage modes with demand-assignment multiple access the estimated capacity of *Intelsat IV* is 4000–5000 circuits.

The electric power required to keep this monster operating is obtained from the incident solar energy on 20·5 square metres of solar cells arranged around the periphery of the cylindrical main body of the satellite. The design capability of this array is 569 watts at the beginning of life and 460 watts at the end of life seven years after launch. Two 15 ampere-hour nickel–cadmium batteries supply power for full operation of the system while the satellite is in the earth's shadow.

Intelsat network improvements

Advances in spacecraft technology in the slightly less than six years between the launching of *Early Bird* and that of the first *Intelsat IV* have made it possible to increase the traffic-carrying capacity by 20 times and to extend the projected life by a factor of 6! Remarkable as these achievements may be, they are probably no more interesting or important than some of the advances in basic communication technology, which have been applied mainly to the earth segment of the system and which have had a great impact on the overall efficiency and flexibility of the system. These advances include the use of more efficient modulation techniques, adaptive multiple-access facilities, and signal processing.

Before we go any further with our discussion of these techniques, it is necessary to define a few of the terms that will be used. First, a radio 'carrier' is the signal transmitted by a radio transmitter. Before being transmitted, the carrier is 'modulated' by varying its amplitude or, more commonly nowadays, its frequency or phase. It is this modulation which contains the information to be transmitted. At the receiver, the carrier is 'demodulated', that is, the information is stripped off and used. Usually a number of telephone or teletype circuits are transmitted simultaneously over a single carrier. Putting them together, which can be done in different ways, is called 'multiplexing'. Taking them apart again is known as 'demultiplexing'. A 'transponder' is the combination of electronic circuits that amplify a received signal or complex of signals and change its carrier frequency with no essential change in information content before it is retransmitted, in this case, back to earth. When a carrier from an earth station can be received and retransmitted properly by a transponder on a satellite, the earth station is said to have 'access' to the satellite or to have 'accessed' it. A satellite that can provide access for several earth stations at the same time has 'multiple-access' capability.

Early Bird was not designed as a multiple-access satellite; its two transponders were 'hard limiting', which means that only one earth station at a time could have access to a transponder. Thus the terminal at Andover, Maine, in the United States, used the west–east transponder on *Early Bird*, while the stations in France, England, Germany, and Italy took turns using the east–west transponder on a weekly basis, except during unusual events. The transponders carried by *Intelsat II*, *III*, and *IV* satellites operate in a 'quasi-linear' mode; several earth stations can therefore use a satellite transponder simultaneously if the power allocated to each carrier is carefully controlled. In this mode of operation, known as frequency-division multiple access (FDMA), each station is allocated a fixed frequency, bandwidth, and power budget. Each potential receiver of traffic from that station must receive its carrier after retransmission by the satellite, demodulate it, and demultiplex the channels of interest for its location. A large station communicating regularly with, let us say, 50 other stations requires a receiver 'chain' for each link and a large quantity of multiplex and demultiplex equipment to process these channels.

SPADE

It soon became evident that *Intelsat* would require new techniques to achieve its goal of rapid global interconnection for all users at a continually decreasing cost. A method known as SPADE (a super-acronym for single channel per carrier pulse code modulation multiple access demand assignment equipment) is one of the answers that has evolved. Some of its key features are as follows.

- Capacity is allocated in units of a single voice channel. This provides necessary flexibility to use the network efficiently when there are many potential users with only a small amount of traffic from each, in other words a 'thin-routes' situation.
- Capacity is shared by all users. When not in use, a channel may be assigned to any user on demand. Each earth-station owner is charged only for the amount of satellite capacity he actually uses. Based on statistics of call placement during busy periods, it is estimated that a satellite capacity of 800 channels can service 2400–3200 individual trunk lines.
- Simple arrangements incorporated in the on-board control equipment make it possible for any calling station to 'seize' available channels automatically—that is, without any action by a central control station.
- Efficient voice encoding and modulation techniques, specifically pulse code modulation (PCM) with phase-shift keying (PSK), are used to maximize the transponder capacity. The encoding of voice or other analog signals into bits (binary digits) has a number of advantages which will be discussed further on in this section.
- SPADE channels all feature voice-activated carriers. This means that unless a party is speaking no satellite power is used. During an ordinary two-way conversation, this feature provides a 60 per cent reduction in radiated power.

As of April 1974, SPADE has been made operational on *Intelsat IV* in the Atlantic region, and its use is growing steadily. Some of the countries using it are now carrying traffic not previously projected to be economic. Thus SPADE seems to be a good solution for the 'thin' routes of a network.

TDMA

However, there still appears to be a requirement for more efficient techniques for the medium-to-heavy traffic routes, and the most promising approach currently seems to be time-division multiple access or TDMA.

The concept on which TDMA is based is that of giving each ground station an assigned time slot for transmitting and receiving. These slots are extremely short in duration but recur frequently and are controlled by a precision crystal clock. The main advantages of TDMA are that it requires only one set of ground equipment to enable a station to communicate with several other stations and that it makes more efficient use of the satellite than other methods because whenever a station accesses the transponder with its timed burst, it does not share the power with any other carrier and thus can operate the transponder close to saturation, where it is most efficient. TDMA inherently requires the use of high-speed digital electronic equipment; present systems can operate in excess of 100 megabits per second. Field trials of TDMA for *Intelsat* are planned for 1975–6, and the Canadian satellite communication system began to use TDMA on a limited basis in 1975.

PCM

Both of the above communication methods provide efficient multiple access for the kind of routes to which they are best suited and both offer additional capacity improvement using signal-processing techniques that are easily implemented because of the digital format of the signals. The first proposal to convert analog signals into digital format was made by A. H. Reeves in 1938. However, it was not until after the invention of the transistor in 1948 that development of a fully successful pulse code modulation (PCM) system became possible. First commercial use of PCM for telephone circuits began in 1962 with the Bell Telephone T1 carrier system; several hundred thousand PCM links are now in operation in the United States, Europe, and Japan. The principal advantage is that use of a digital format allows transmission of information over long distances or noisy circuits without deterioration because digital signals, unlike analog signals, can be regenerated with only a very small probability of error. Many other problems, such as electronic switching and cross talk, are much easier to handle with digital signals, although some new problems are introduced.

In a typical PCM system, the audio signal from a subscriber's telephone is passed through a low-pass filter which eliminates all frequencies above four kilohertz. This filtered signal is then sampled at a rate of 8000 times each second and each sample is converted into an eight-bit binary number by a high-speed analog-to-digital converter. The same converter is also used on a time-shared basis with up to 29 other links, so its output is a digital data stream containing a maximum of $8000 \times 8 \times 30$ or 1·92 megabits per second. This data stream is sent via satellite, terrestrial microwave relay, coaxial cable, or other broad-band transmission system to a digital-to-analog converter and demultiplexer at the other end, where it is reconstituted into 30 separate simultaneous audio signals and routed to the individual subscribers. (The specific parameters given here are typical, but are for illustration only.)

Delta modulation

A newer and better method of digitally encoding voice signals, known as delta modulation, was first developed in France about 1946. In delta modulation, an analog signal is synthesized by identical electronic circuits at both ends of the transmission link. At the transmitting end this reconstructed signal is compared at a clock-controlled sampling rate with the original input signal. If there is any error, the sign (+ or −) of that error is converted into a digital 1 or 0, which is sent over the transmission system and used to cause the reconstructed signal at both ends to be changed by one unit in the correct direction. If the reconstructed signal still differs from the real signal in the same direction, the same binary number will be transmitted, again and again if necessary, until the difference (error) changes sign. When the reconstructed signal and the real input signal are essentially equal 1s and 0s will be transmitted alternately. Because only a single binary number or bit is transmitted at each sampling point, instead of a seven- or eight-bit number as in PCM, the sampling frequency can be several times greater and still keep the total bit rate lower than that for PCM. An alternative form of delta modulation compares the slope of the reconstructed signal with that of the input signal, and encodes the sign of this difference in slope as a binary 1 or 0. This method lacks inherent stability and can oscillate under certain conditions. Consequently a combination of the two methods is generally used that gives superior performance and yet retains good stability.

Delta modulation, unlike ordinary PCM, requires no expensive audio filters with steep-slope characteristics. It does have a problem with dynamic range, however. Low-intensity voice signals are quantized rather roughly compared with PCM. The most common solution to this problem is 'companding', that is, *com*pressing the signal dynamic range at the transmitting end and ex*panding* it at the receiving end. Hi-fi enthusiasts will be familiar with this process.

The transmitting of digital data

With the introduction of SPADE operations in *Intelsat* it became evident that, since individual voice channels were being transmitted digitally at 64 kilobits per second, it would be quite practical to transmit digital data at similar rates. One drawback existed. The voice channels operate at a threshold bit error rate of about 1 in 10 000, which provides good voice quality but is not acceptable for many data applications. To overcome this drawback error-correcting coders are used which require that some additional bits be transmitted but which substantially reduce the error rate. Detailed descriptions of these codes and coders can be found elsewhere, but as an example, if a stream of 48–50 kilobits per second of data is passed through a 3/4 convolutional encoder, the net channel rate is then 64–67·7 kilobits per second, which is compatible with SPADE capability. The resultant bit error rate after reception and decoding is not more than 1 in 10^8 at the 48–50 kilobits per second input–output rate. Several data channels of this sort are already in operation across the Atlantic.

Domestic satellite systems

Intelsat satellites, though primarily intended for international service, can of course be used for communication from one part of a large country to another. Australia, for example, has used them in this way. Canada has established its own satellite system for domestic communications, known as the *Telesat* system. Its first satellite, called *Anik* (an Eskimo word for brother), was launched in November 1972. Five or six domestic-system operators in the United States have also been authorized to move ahead with their initial plans. France and Germany have a co-operative programme called *Symphonie* for European and North African coverage. Japan is planning both a domestic communication satellite and a television broadcast satellite, of which more later. It seems suddenly to have become very important for each country and each user to plan for its own domestic satellite communication network. Economic restraints will undoubtedly put finish to some of these plans.

The Soviet Union also has been active in this field. Her *Molniya* communication satellites are placed in highly elliptic orbits with an apogee of about 40 000 kilometres in the northern hemisphere, 64° inclination, and a 12-hour period of revolution. Three of these satellites appropriately phased in time can provide round-the-clock service over the vast European and Asian areas of the Soviet Union without the additional complexity and cost of putting satellites into geostationary orbits from the high northerly latitudes of the Soviet launch pads. The *Molniya 2* satellites have several transponders operating in the 6 gigahertz and 4 gigahertz bands, at least one of which is designated to provide single channel per carrier (SCPC) capability. The *Molniya* system is not a purely domestic system. It provides regular service to other socialist countries in eastern Europe and also has provided communication service at least part time with Cuba. Plans are now being implemented to provide two direct satellite communication links between the United States and the Soviet Union for use by the Heads of State, one through the *Intelsat* system and one through *Molniya*.

Military satellite communication systems

Although the first artificial earth satellites were developed for non-military, scientific purposes, it did not take long for the military services to understand the value that satellite communication systems could have in strategic and tactical operations and in the integration and supply of their far-flung forces and facilities. Early experiments, in particular the *Advent* programme, were useful in developing concepts, technology, and components, but were not successful operationally because they were too ambitious for the then-available launch vehicles and spacecraft technology.

The first operationally useful system was the Interim Defense Communication Satellite Project (IDCSP), later renamed Defense Communications Satellite System, Phase I (DSCS-1). This system made use of a total of 26 satellites, launched between June 1966 and June 1968. These were simple spin-stabilized spacecraft with no moving parts, designed for the highest possible reliability. They were deployed in slightly sub-synchronous orbits with an eastward drift of about 26° per day, in groups of as many as seven at a time, using a *Titan IIIC* (large military rocket) launcher in a complicated series of manoeuvres requiring repeated restarts of the transtage rocket motor. Although no station-keeping capability was provided, each satellite remained in view of an equatorial station for four or five days, and several were usually available at any given time. Limiting wide-band FM transponders were used in the seven to eight gigahertz band. Some 36 earth terminals (including seven for ships) were constructed for this interim system. The DSCS-1 was capable of linking stations up to 16 000 kilometres apart on the earth's surface. South Vietnam–Hawaii–Washington links were used for high-speed digital data and high-quality reconnaisance photographs during the last seven years of American involvement in the Far East. As of spring 1974, 11 of the 26 satellites were still operating satisfactorily

The technology developed for the DSCS-1 with some improvements was used in two other military satellite communication systems, one for the United Kingdom, called *Skynet*, and the other for NATO. The *Skynet* system included large fixed ground stations and shipboard as well as land mobile (or transportable) terminals. Two satellites were built in the United States and launched for the United Kingdom under a co-operative programme in 1969 and 1970. Only one was successful. Two other satellites built in the United States were successfully launched for NATO in 1970 and 1971. The operational system includes 12 ground stations located near the capitals of the 12 NATO countries. All four of these spacecraft are of essentially the same design. The transmitter power is only slightly greater than that of DSCS-1, but a mechanically despun antenna provides enough gain for the signal at a ground station to be more than an order of magnitude greater. They are placed in geostationary orbits and have station-keeping capability.

A *Skynet 2* programme, using satellites of an improved design built in the United Kingdom, was initiated in 1971. The first *Skynet 2* satellite launching by the United States in January 1974 was unsuccessful, but the second attempt in November 1974 succeeded. An advanced phase of the NATO programme has also been initiated that will increase the number of ground stations to 22, add two large transportable terminals, and extend service to ships under NATO command. The satellite for this NATO Phase 3 will use two transponders, one connected to a wide-beam antenna, and the other to a narrow-beam antenna providing high-gain coverage only over Europe.

Fig. 4.2. DSCS 2 satellite being tested in a special chamber the walls of which do not reflect radio waves.

Fig. 4.3. *Tacsat 1* satellite in comparison with *Syncom*, the first synchronous communication satellite.

The follow-on satellite for the DSCS system, known as DSCS-2, has multiple-channel, wide-band, adjustable-gain transponders with one earth-coverage antenna and two narrow-beam (2·5°) steerable antennas. It can more readily link the smaller ground terminals now under development. The first two of these satellites, one of which is shown in Fig. 4.2, were launched late in 1971 and two more were put in orbit in 1973. They are now operational.

All of the above military satellite developments have been intended primarily for long-haul strategic command and control and for defense management communications. Largely independent of this effort, the United States Department of Defense has been investigating the potential of satellites for mobile tactical communications which require very small lightweight earth terminals and preferably a capability to use existing operational UHF bands. This capability was originally demonstrated by the Lincoln Experimental Satellite LES-5, launched in 1967, and the LES-6, launched in 1968. The *Tacsat* programme is an outgrowth of this effort and has provided an operational capability since 1970.

The *Tacsat 1* satellite, launched into a geostationary orbit in 1969, is the Western world's largest communication satellite. Shown in Fig. 4.3 it dwarfs the *Syncom* satellite launched six years earlier. It must be big, of course, because it must work with a multiplicity of very small ground terminals. Stabilized by spinning its outer shell, it is equipped with a large despun platform that mounts three accurately pointable antenna systems. One of these consists of five helical UHF antennas; the other two are microwave horns for X-band communication and for telemetering and command. Communication capability of the spacecraft is said to be comparable to 10 000 two-way telephone channels. New terminals are being tested on jet aircraft, small ships, jeeps, and even back-packed equipment. It would seem that a great advance has been made since the telephone wires of the First World War and the crackly, undependable HF radios of the Second World War!

Further development of mobile services

The military tactical communication capability will be extended by the United States Navy when the *FleetSatCom* satellite is launched in 1976. Meanwhile, *Comsat* General, in conjunction with other American communication organizations, is planning to establish a combined commercial and military maritime satellite system called *Marisat.* Two satellites, one over the Atlantic and one over the Pacific, will provide ship-to-ship and ship-to-shore service in the VHF band (240–400 megahertz) and in UHF bands near 1600 megahertz. Shipboard antennas will be about 1·25 metres in diameter. The United States Navy has arranged to lease some of the UHF channels on *Marisat* until *FleetSatCom* becomes operational. On the European side, the European Space Research Organization (ESRO) has announced its intent to launch its own experimental maritime satellite with the name *Marots.* Ultimately, sometime in the 1980s, the International Maritime Consultative Organization (IMCO) intends to establish a fully internationalized maritime satellite communication service.

There is clearly a need for a dependable means of communicating with commercial airliners flying over the oceans, out of line-of-sight range with shore stations. A joint venture, involving participation by the United States, Canadian, and European entities, will fill this need beginning in about 1977 with a system known as *Aerosat.* This system also will include one satellite over the Atlantic and one over the Pacific. Communication channels will be provided at both VHF and UHF frequencies, and

Fig. 4.4. ATS-F satellite being tested at Fairchild Industries before shipment to NASA Kennedy Space Center. Large parabolic reflector and two half cylinders covered with photoelectric cells are the most prominent parts. Science experiment package is at the top, and the earth-viewing module, which contains electronics, is at bottom.

it is anticipated that these satellites may be used for position-fixing as well as for communications.

Experimental communication and television broadcasting from satellites

Beginning in 1966, NASA designed and constructed a series of experimental satellites for trying out various useful applications of satellite technology including communications. These have been known as the Applications Technology Satellites (ATS), and there have been six of them, of which three were successfully placed in geostationary orbits. The most recent was ATS F (Fig. 4.4), which was launched in the spring of 1974. It is something of a giant, being the first satellite capable of deploying a 9·1-metre parabolic radio reflector in space. In addition to other experiments it can broadcast television at 2·5 gigahertz direct to small individual receiving terminals in Alaska and other remote areas of the United States and also at 860 megahertz for a domestic educational system experiment in India. The Brazilian Government is studying the use of the 2·5 gigahertz channel for educational broadcasts to serve its various wilderness communities. The transmitters on ATS F operate at about 15 watts. The ground terminals therefore require not only a receiver 'front end' for the right frequency and modulation system but also some additional antenna gain, such as might be supplied by about a two-metre diameter reflector.

In other countries, planning for broadcast satellites has moved toward the 12 gigahertz band rather than the 2·5 gigahertz band. The Japanese Government has already placed a contract for development of a 12-gigahertz broadcast satellite using a 100-watt travelling-wave-tube transmitter, suitable for operating with very small ground-terminal antennas. The Canadian Government is building a new Canadian Technology Satellite, with a transponder that will use a 200-watt travelling-wave tube at 12–14 gigahertz. The Federal Republic of Germany has made a significant study of a 12-gigahertz satellite for broadcasting directly into the home, using a 700-watt travelling-wave tube which is being developed in Germany.

The concept of direct broadcast satellites is one which can stir up vigorous controversy in political circles. Because of its potential for use as a propaganda medium, some kind of international control will almost certainly be proposed. On the other hand, because of its potential for education in disadvantaged areas and the benefits in terms of improved agriculture, health care, and population control which will flow from this potential, direct broadcasting will surely be implemented in at least some areas.

For good or evil?

The image which comes to mind of tens of thousands of people all talking at once across continents and oceans, of high-quality colour television pictures of football games bouncing from one side of the earth to the other, and even of great electronic computers communicating with each other at speeds vastly beyond our power to comprehend may make one wish for simpler days when one could only send a letter and wait for the reply. But those days will not come back again. The means to communicate is a commodity which is clearly desired by people all over the world. The amount of this commodity that can be sold at prices high enough to cover the cost is still increasing, and no ultimate limit can yet be defined.

Whether this new capability will serve for the benefit or the detriment of mankind will depend, of course, on how mankind will choose to use it. In a sense, perhaps we are seeing a reverse of the Biblical story of the tower of Babel (Genesis 11). The reader may recall that after the great flood, 'the whole earth had one language and few words.' The people set to work to build a city and a tower with its top in the heavens in order to 'make a name for ourselves, lest we be scattered abroad upon the face of the earth'. But the Lord saw what they were doing and He said 'Behold they are one people and they have all one language and this is only the beginning of what they will do and nothing that they propose to do will be impossible for them.' So He confused their language and scattered them and they 'left off building the city'.

Perhaps now that people can put platforms even higher than the heavens and use them to communicate with each other, some of the confusion will be eliminated and once again nothing that they propose to do will be impossible for them—even to coexist peacefully as one family on the earth.

5 Watching the weather

Sometimes wind and sometimes rain,
Then the sun comes back again,
Sometimes rain and sometimes snow,
Goodness, how we'd like to know
Why the weather alters so.

Ford Madox Ford (1873-1939) *Children's song,* Stanza I

Probably the most profound statement that can be made about the weather is simply that it changes. Scientists and weather forecasters would surely like to know much more than they do about how and why it changes—to know enough to predict future changes accurately and to cause desirable changes to happen, if indeed most people could ever agree on what changes would be desirable.

Even in this age of central heating and air conditioning, the weather affects everybody. A few really warm days in early spring, for example, send air-conditioner sales soaring and make jobs for people who manufacture air conditioners; the added air conditioners may then overload the electric utility system later on in the summer and cause power failures. If a sudden cold spell hits Florida, half the orange crop becomes frostbitten and worthless. If a cold spell with plenty of snow hits the ski areas of Vermont or the Alps just before the holidays, the hotels and ski-lift operators have a good year. If a hurricane roars up the Gulf of Mexico it flattens houses and floods the coastal area where it strikes; then it dissipates over the inland area beyond, dumping its load of millions of tons of rain and leaving a trail of death and destruction behind. Or if the monsoon rains fail to develop properly in India, rice crops fall, and people starve.

A brief history of meteorology

From the beginning of civilization a host of deities have been imagined in an effort to explain phenomena of nature for which no rational explanation could otherwise be devised. Thus for many centuries, men believed that the weather changed at the whim of the gods. Bad weather, especially, was one of the ways in which the gods showed their displeasure, and storms were interpreted as cosmic struggles between these immortal beings. However, by the fifth century B.C., although the Greeks were still attributing weather to their gods and demigods, they were also publicly posting wind observations for mariners. In the fourth century B.C., Aristotle wrote a treatise on the physics of the earth and the air, which he called *Meteorologica.* As was typical of Aristotle, he attempted to arrive at scientific truth by logic rather than by detailed observation; thus his ideas about the weather were hardly less imaginative than some earlier theistic concepts. It was an important step, nevertheless, because it established with considerable authority that weather was subject to the laws of nature rather than the whims of the gods. Aristotle's pupil Theophrastus in his *Book of signs* compiled some 200 portents of fair and foul weather based on a wide variety of common observations such as the behaviour of animals and insects and the burning of a lamp. Some of these are still considered to be valid, such as the probability of rain after a red sunrise or the appearance of a halo around the sun or moon.

The authority of Aristotle, Theophrastus, and the old shepherds tales continued for almost 2000 years, in parallel with the rather firm belief that God could change the weather to benefit or punish his followers if He wanted to. Only during the Renaissance—when, under the influence of Galileo, Torricelli, Hooke, and others, accurate instruments were invented for making meteorological measurements and Newton was establishing the underlying principles or 'laws' of dynamics—did a truly scientific study of weather begin.

In the years that followed there were many important steps forward in meteorology, of which two seem to be of particular significance. One was the development of the balloon in 1783, which enabled scientists to carry their instruments up into the weather and thus to obtain a better understanding of its three-dimensionality. The other was the invention of the telegraph in 1844, and its almost immediate application to the regular transmission of weather observations. In the United States, early telegraph operators were provided with meteorological instruments by the Smithsonian Institution, in return for which they made observations and transmitted them to Washington, where current weather maps were compiled and published. Unfortunately the Civil War in the United States broke the links of this weather-reporting system and it was never re-established. Government weather forecasting finally began as a storm-warning system for Great Lakes shippers in 1870 with a telegraphic network of 25 observing stations under the control of the United States Army Signal Service.

The scientific and utilitarian aspects of meteorology began to come together around 1900 through the efforts of a group of Norwegian scientists lead by Vilhelm Bjerknes. By this time, weather data had become more accurate and more comprehensive so it seemed reasonable to suggest that forecasting could be transformed from educated guesswork to something like an exact science. Bjerknes believed that the existing state of the atmosphere as recorded by a network of data-collecting stations could be extrapolated by means of mathematical equations to compute the weather to come. He never completely achieved this goal but, with the help of the Norwegian fishing-fleet operators, he established a world-famous weather service and meteorological institute in Norway and developed a useful theory relating the motion of polar and mid-latitude air masses to the formation of large-scale cyclonic disturbances.

Meanwhile a British mathematician named Richardson thought it might be possible to solve the basic equations describing the global motion of the atmosphere by a numerical step-by-step procedure using mechanical calculating machines. However, he concluded that the task would require 2000 permanent weather stations, equipped for both surface and upper atmosphere observations, and 64 000 calculators operating 24 hours a day, seven days a week. The 'forecast factory' he described was pure science fiction at the time (1922), as even he realized.

In the period since the Second World War, four major new technological developments have been brought to bear on the problems of meteorology. One is the use of radar for extending the 'vision' of individual weather stations, enabling them to map precipitation to a range of a hundred kilometres or more. The second is the radar-tracked balloon sonde that can routinely and accurately measure winds as well as other meteorological parameters at various levels in the atmosphere. The third is the use of satellites for making weather observations all over the earth. Much more will be said about this later. And finally there is the availability of electronic digital computers that can do the work of more than 100 000 calculators, at a cost that is within the reach of any large government weather service. Unfortunately, poor Richardson had his vision some 40 years too soon!

Forecasting

Weather forecasting has two aspects. One is the largely descriptive phenomenological aspect, which has developed out of the accumulated experience of many expert meteorologists studying many thousands of weather maps, guided by an increasing body of scientific understanding of the physics of the atmosphere. The other is the analytical computational aspect, which was anticipated by Bjerknes and Richardson and is now becoming feasible because of the developments in electronic computer technology.

The first was described by a former Navy weatherman in the following terms: 'It is as simple as it sounds. We just analyse the large bulks of air from reports on the weather maps, assign each type a name, separate them by fronts, find out where they have been and where they are going, and make forecasts accordingly'. (Donald A. Whelpley (1961), *Weather, water and boating*) It does sound simple. What has been omitted is that it is also necessary to understand how one air mass interacts with another at the front that separates them, how low and high pressure areas form, intensify, combine, and dissipate, how the speed with which fronts advance varies with season and situation, what happens when one front overtakes another, how geographical features such as mountains, large lakes, shorelines, and islands may affect the weather features, and much more. Even then, such forecasts are only for short time periods, usually of 24 hours, or at most 48 hours, and the accuracy usually leaves something to be desired.

Fig. 5.1. Weather features over the United States compared with satellite cloud picture, 18.00 GMT, 8 August 1972.

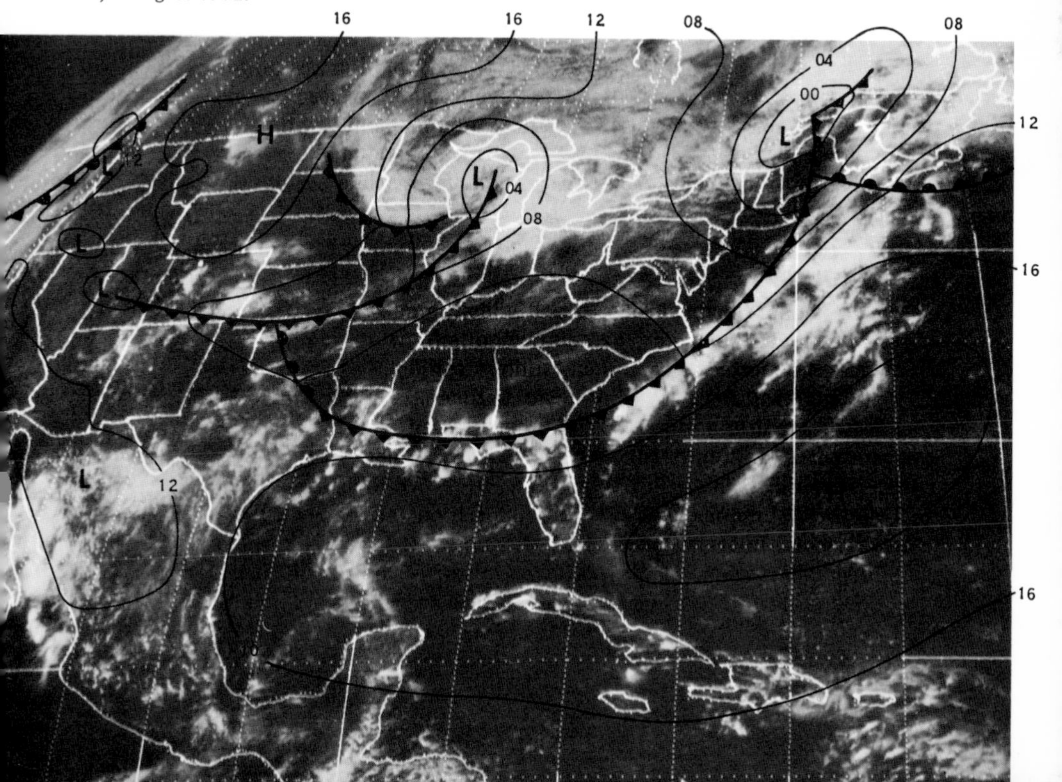

Local forecasts

Forecasting of this kind is primarily local and regional in character. In order to be effective it does need good inputs in the form of observations and weather maps, especially covering the area from which weather features generally move into the region being forecast, usually to the westward in the temperate zones. This is one area where satellite cloud pictures can be helpful, for the pattern of clouds in such pictures can usually be interpreted by an experienced meteorologist to determine or verify the position and character of fronts and low-pressure areas with a fair degree of accuracy. Fig. 5.1 shows how this can be done. Modern weather satellites are equipped with an automatic picture transmission (APT) system which enables the local or regional forecaster to receive directly from the satellite a high quality picture of the clouds in his area. These APT pictures which can be received with low-cost equipment can be of enormous value to the weather services of small countries, especially in parts of the world where the availability of conventional observations to the westward is not good. They are also the source of the satellite pictures frequently shown by television weathermen on their evening broadcasts.

Worldwide forecasting

The other aspect of weather forecasting involves the creation of a mathematical model of the atmosphere based on the fundamental equations of motion, continuity, and thermodynamics. With initial values established by an observational network, the motion and associated changes in temperature, pressure, and other properties of meteorological interest are computed, step-by-step, for successive small time intervals until the end of the desired forecast period is reached. A forecast is then made, and the data in the computer is modified to take account of new observations. This kind of forecasting is not nearly as simple as it sounds. The model cannot ever be a perfect representation of the real physical situation and to be of value it must be designed to be able to be run on existing computers in much less than real time (the time required for the changes to occur in nature). Obviously a 24-hour weather forecast that would require more than 24 hours to produce would be worthless. Thus with even the best of modern computers the model cannot take into account many important phenomena nor compute in as much spatial or temporal detail as would be desired. These shortcomings sometimes introduce computational instabilities and other difficulties that are gradually being overcome but at the expense of added complexity and poorer forecasting accuracy.

Weather forecasting of this kind must be done on a worldwide scale for there are no natural boundaries of latitude or longitude across which atmospheric variations cannot penetrate. A typical model may require 10 000–20 000 machine instructions and may have to compute temperature, pressure, and humidity as well as wind speed and direction, for as many as 10 levels in the atmosphere at each of about 10 000 grid points over the surface of the globe, a total of about half a million variables. From various tests which have been made it is expected that a model of this complexity should be able to produce useful forecasts of synoptic (large-scale) weather features for a period of one to two weeks ahead if adequate input data can be provided and if certain physical relationships embodied in the model (such as those relating to momentum, heat, and water-vapour exchange in the lower boundary layer and those relating to cloud convection) can be better understood and more accurately defined. A major international effort, known as the Global Atmospheric Research Programme, is now underway to accomplish these objectives.

Meteorological satellites will be essential for this second kind of weather forecasting. They can measure temperature and humidity profiles all over the earth, from which pressure fields can be determined and winds calculated, at least in non-equatorial regions. They can also measure winds directly at high altitude by tracking constant-pressure balloons, at lower altitudes by following the short-term motion of individual cloud patterns, and at sea level by relaying information from buoys or by observing the height and wavelength of ocean waves. They can measure the height of cloud tops, approximate their water content, and measure the surface temperature of the oceans very accurately under cloudless skies and less accurately under clouds. They can detect the formation of intense tropical storms and provide reasonable estimates of the wind fields within and surrounding them at various stages in their development. By no other means can this wealth of data which is necessary to the success of numerical weather forecasting be obtained.

Observation of cloud systems and precipitation

Cloud pictures can be made from satellites by two basically different methods. In one, an image of the clouds is focused on the sensitive surface of a television tube, usually a vidicon tube, and scanned electronically. In the other method a radiometer (an instrument which converts the intensity of electromagnetic radiation falling on it into a proportional electrical signal) is combined with an optical system so that it is sensitive only to radiation coming from a single narrowly defined direction, and this direction is scanned over the clouds by means of a moving mirror. The earliest cloud-imaging satellite experiment, developed by Stroud and flown as part of the International Geophysical Year programme, was of the second type. It depended on the motion of the satellite for scanning, and a special machine was built to convert the video signal which it sent back to earth into a cloud image. Unfortunately, during the separation of the satellite from its launcher, it was nudged in such a way that its subsequent motion was quite different from the intended motion so the data reduction machine was useless. Converting the data into images manually proved to be so expensive that only a small amount of it was ever processed. A few years later as satellites became larger and could provide more electrical power a vidicon system was flown which produced images of good quality that clearly demonstrated how useful such pictures could be.

Vidicon systems

The present state of the art is exemplified by the advanced vidicon camera subsystem (AVCS) which has been flown first on *Nimbus* experimental weather satellites and later on operational *Tiros* satellites. A photograph of this instrument is shown in Fig. 5.2. Three separate cameras are used, one pointing straight down and the other two tilted outward by 35°, perpendicular to the direction of flight. The angular size of each image is 37° square. With a 2° overlap, therefore, the three cameras simultaneously view an angle of 37° X 107°. From an orbit at 1130 kilometres altitude, the AVCS can cover a strip about 3600 kilometres wide around the whole earth during each orbit. This is wide enough for the entire surface of the earth without any gaps even at the equator to be viewed twice each day, once in daylight and once at night, by a polar-orbiting satellite. If a sun-synchronous orbit (see p. 13) is

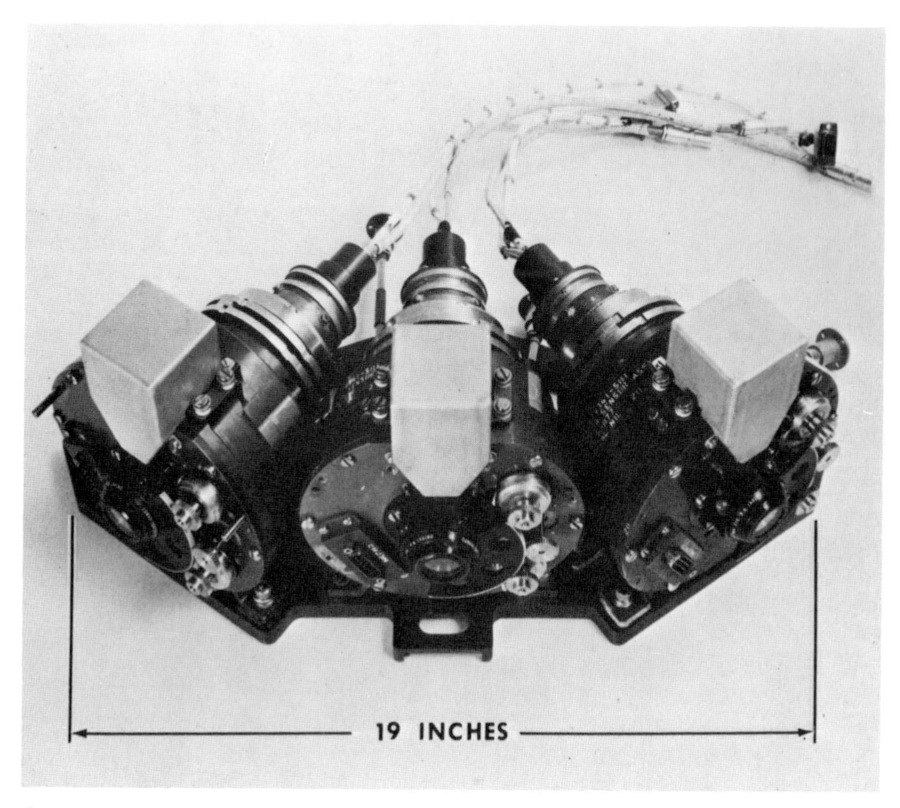

Fig. 5.2. AVCS cloud-camera subsystem as flown on *Nimbus* weather satellites.

used, as is customary for low-altitude weather satellites, the satellite will pass over all points at about the same local time on a 12-hour clock. The time can be selected to suit the needs of the weather forecaster. Obviously the AVCS, which uses visible light, cannot 'see' in the dark, so it is switched off on the night side of the earth and other instruments that use infrared radiation, of which more will be said later, provide the cloud pictures at night.

The resolution of the AVCS at nadir (looking straight down) is about 1 kilometre. At least six shades of grey can be registered in the central portion of the grey scale. Fig. 5.3 is a mosaic of AVCS pictures showing hurricane Alma over Tampa, Florida, on 9 June 1966.

The automatic picture transmission (APT) subsystem is another vidicon system. It is intended to take wide-angle pictures of clouds continuously during daylight hours and transmit them in real time to any ground station in the area for reproduction on ordinary facsimile equipment. A wide-angle lens makes it possible for one camera to cover an area about 2320 kilometres on a side in each frame, and frame speed is adjusted so that there is about 550 kilometres overlap between successive pictures. A ground station can usually record two or three such pictures on every pass of the satellite over its location. Resolution under ideal conditions is about 4·4 kilometres in the centre of the picture. Fig. 5.4 is a high-quality photo-

Fig. 5.3. Mosaic of AVCS pictures showing hurricane Alma over Tampa, Florida, on 9 June 1966.

Fig. 5.4. APT image of same storm shown in Fig. 5.3.

graphic reproduction of an APT picture of the same storm shown in Fig. 5.3. The principle advantage of this system is that any weather service can receive pictures of its own local weather with essentially no time delay using very-low-cost equipment.

Scanning radiometer techniques

Despite the success of the vidicon approach to cloud imaging, there is now a trend in the opposite direction, back toward the scanning radiometer technique. Infrared sensors have proved to be very useful because they respond to thermal radiation rather than to scattered or reflected sunlight. They can therefore operate at night as well as during the daytime, they can measure the temperature of cloud tops and thus their approximate height, and by selecting appropriate bands can map areas of high and low atmospheric humidity, small differences in surface temperature, and other meteorologically interesting properties. Especially important is the observation that, when an infrared channel is used together with a visible radiation channel, areas of snow and ice cover can easily be differentiated from clouds. Vidicon tubes cannot be used at thermal infrared wavelengths; consequently scanning radiometers must be used for infrared imaging, but fortunately they perform quite well. As instruments of this kind can be used to scan simultaneously in several different wavelength bands, thereby producing images which are inherently well registered geometrically with each other, it now seems desirable to include a visible radiation band along with the infrared bands in a single scanning radiometer instrument rather than to carry a separate television camera for the visible part of the spectrum.

Several different scanning radiometer instruments have been used in weather satellites. Typically they make use of reflective optics, including a rotating plane mirror that causes the sensitive direction of the instrument to scan across the earth's surface at right angles to the motion of the spacecraft. Taking the forward speed of the spacecraft into account, this scanning motion results in an image consisting of a continuous strip with a line spacing on the ground of about 0·75 kilometres. The visible and infrared energy collected by the primary optics is split into two or three parts, each of which is passed through a separate band-pass filter and focused on a detector.

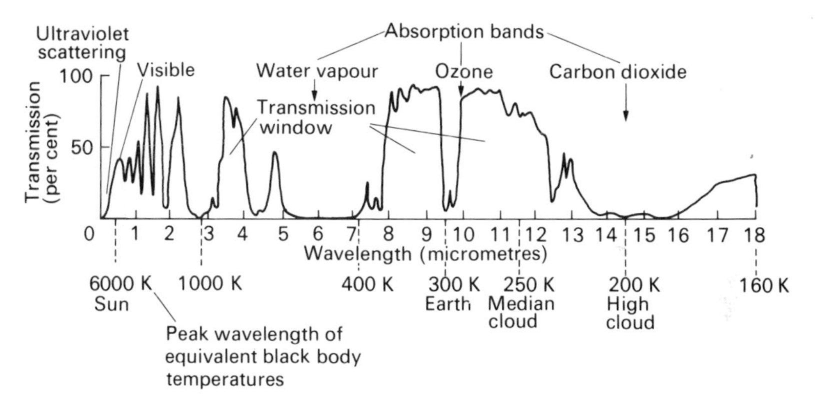

Fig. 5.5. Transmission of the atmosphere at visible and infrared wavelengths.

At least one of the pass bands is usually in an atmospheric transmission window (a part of the spectrum where radiant energy is not significantly absorbed by the atmosphere), such as the region between 10 micrometres and 12 micrometres or the region between 8·5 micrometres and 9·5 micrometres in Fig. 5.5. Most of the energy reaching the detector in this band comes from the surface of the earth or from clouds, so the detector output can be used for surface-temperature mapping or for indicating the height of clouds by means of their temperature. High clouds are generally colder than low clouds. A second pass band is usually located in a strong water-vapour absorption region such as that between 5·5 micrometres and 7 micrometres in Fig. 5.5. This detector will, for the most part, measure infrared energy radiated by water vapour in the atmosphere and its output is therefore useful for mapping atmospheric humidity and high thin cirrus clouds which are difficult to observe by other means. A third band in the visible or near infrared (0·8–1·1 micrometres) is sometimes added for mapping clouds in the daytime by reflected sunlight.

Fig. 5.6. Infrared image made at night over Florida, by SCMR on *Nimbus 5* orbit 173. Shows Florida, Cuba, Bahama Islands, and tip of Yucatan Peninsula.

Fig. 5.6 illustrates the excellent detail that can be seen in images made at night by a good scanning infrared radiometer. In this picture one easily sees the state of Florida in the United States and a number of lakes which it contains. The lakes appear to be black because they are warmer at night than the surrounding land. The Bahama Islands, Cuba, and the Yucatan Peninsula can also be identified. The whitest areas are high clouds, because they are very cold. Greyer areas are lower clouds or land masses. Fig. 5.7 shows the difference between an image using the 11·5-micrometre window band (left) and one using the 6·7 micrometre water-vapour absorption band (right). Thin high-level cloudiness can be seen in the water-vapour image that is not evident in the other. However, the land–sea boundary can be made out only in the image from the window band. The scene is a small tropical storm near North West Cape, Australia. Fig. 5.8 is an infrared view during the daytime of a particularly disasterous tornado-spawning storm over the central part of the United States in April 1974. The complex pattern of vorticity and the several cloud levels characteristic of such a storm can be clearly seen.

The cloud mapping instruments described in the foregoing paragraphs have been designed to be used on satellites in low-altitude, sun-synchronous, near-polar orbits and have clearly proved their value. However, there are some drawbacks, one of the most important of which is that a single satellite can view a particular area only once every 12 hours. Weather patterns develop, decay, and change their motion so rapidly that it would be desirable to view them more frequently. Also, determination of winds aloft by observing the motion of particular cloud elements is exceedingly desirable, but this cannot be done unless successive pictures of the same cloud formations taken no more than about 30 minutes apart can be compared. Therefore a device has been developed for taking cloud pictures from geostationary satellites situated about 36 000 kilometres above the earth's equator. The first device of this kind, known as a spin camera, was tested successfully on ATS 1 in 1966. More recently, 17 May 1974, a synchronous meteorological satellite (SMS) was launched with an improved version of this camera. After a period of experimentation it is expected that it and another like it will be stationed over South America and the equatorial Pacific Ocean as the first operational geostationary meteorological satellites under the name GOES, which is an acronym for geostationary environmental satellite. These two satellites together will cover the western hemisphere from northern Canada and Alaska to Antarctica, with overlapping coverage of at least a quarter of of this area. Each satellite will be able to produce and send back to earth a high-resolution cloud picture of about a quarter of the earth's surface every 30 minutes or of a smaller selected area even more frequently.

The imaging device that performs this miracle is called a visible–infrared spin–scan radiometer (VISSR). (The jargon gets worse and worse but really there is no way to avoid it!) In this case, the whole satellite spins at 100 revolutions per minute about an axis parallel to the earth's axis. This motion provides the scanning achieved by the rotating mirror in the instruments previously described. Because the satellite is in a geostationary orbit, however, and does not move with respect to the earth's surface, translation of the satellite cannot be used for the other scanning motion, analogous to the vertical sweep on a television screen. Therefore a mirror is still needed but in this case it is indexed by a small angle about an axis perpendicular to the satellite spin axis once each revolution. A conceptual view of this arrangement is shown in Fig. 5.9. Eight identical photomultipliers scan simultaneously in the vis-

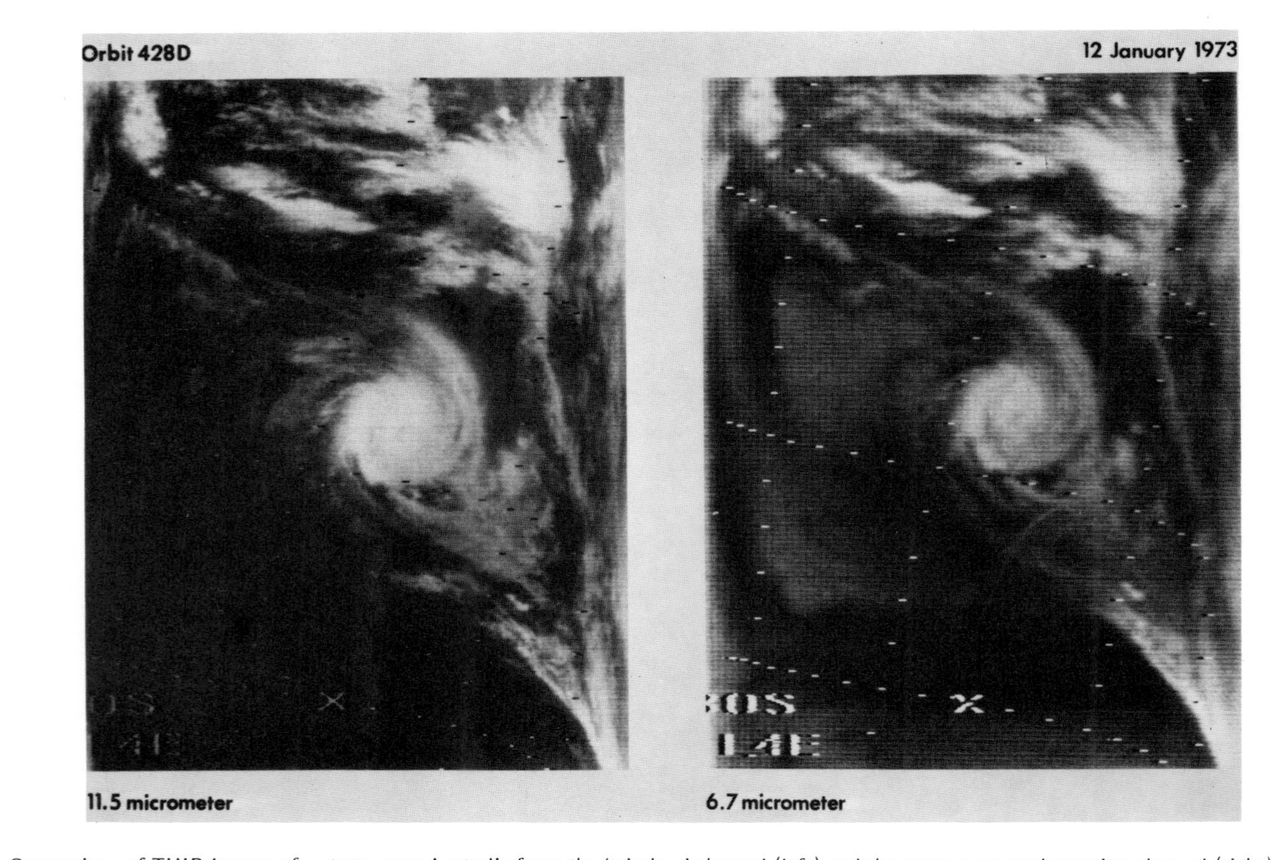

Fig. 5.7. Comparison of THIR images of a storm over Australia from the 'window' channel (left) and the water-vapour absorption channel (right). Recorded during 428D, *Nimbus 5*.

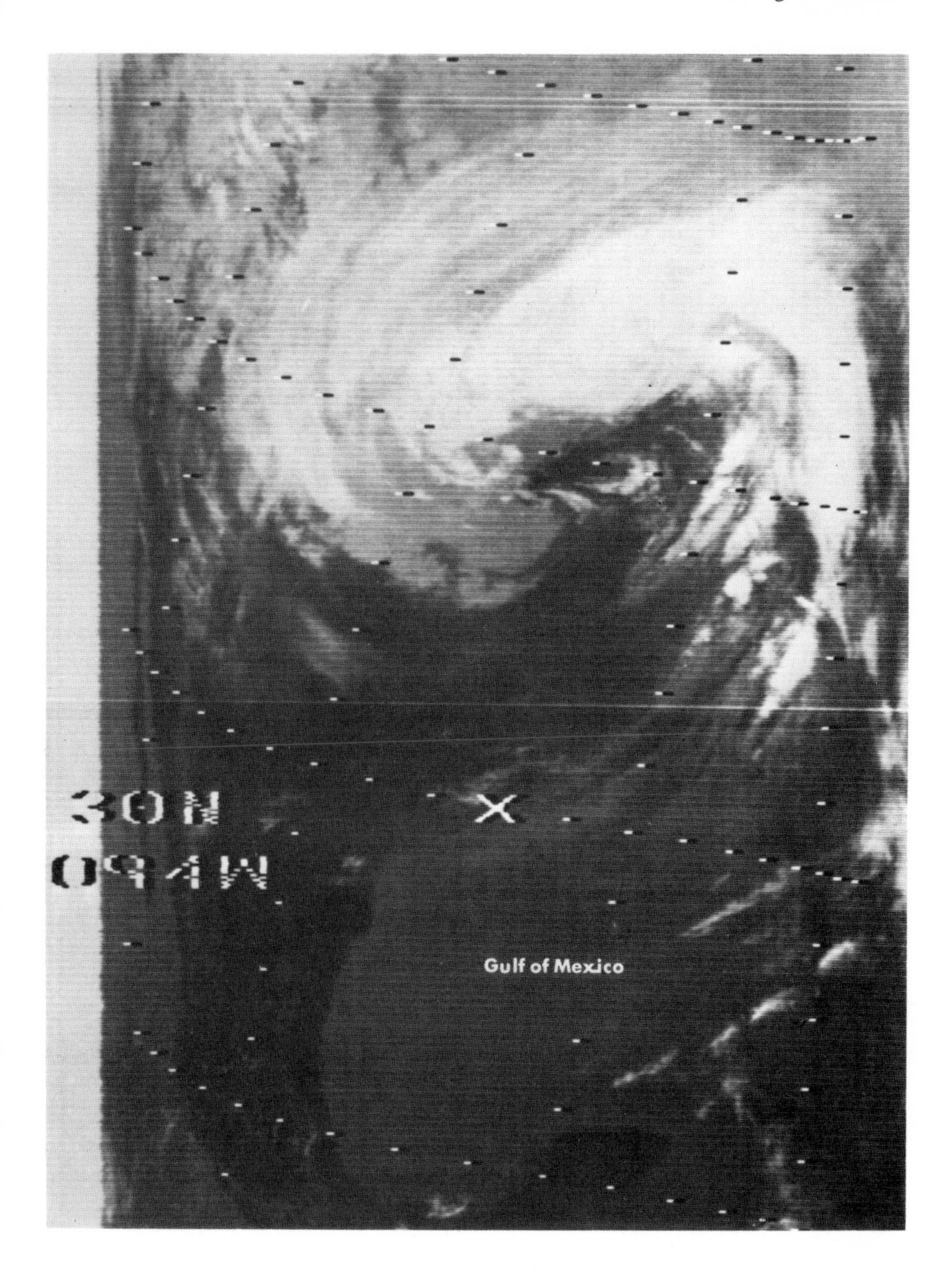

Fig. 5.8. THIR image of storm over United States recorded on *Nimbus 5* at 13.20 on 3 April 1974. A few hours later tornadoes devastated much of the storm area killing more than 300 people.

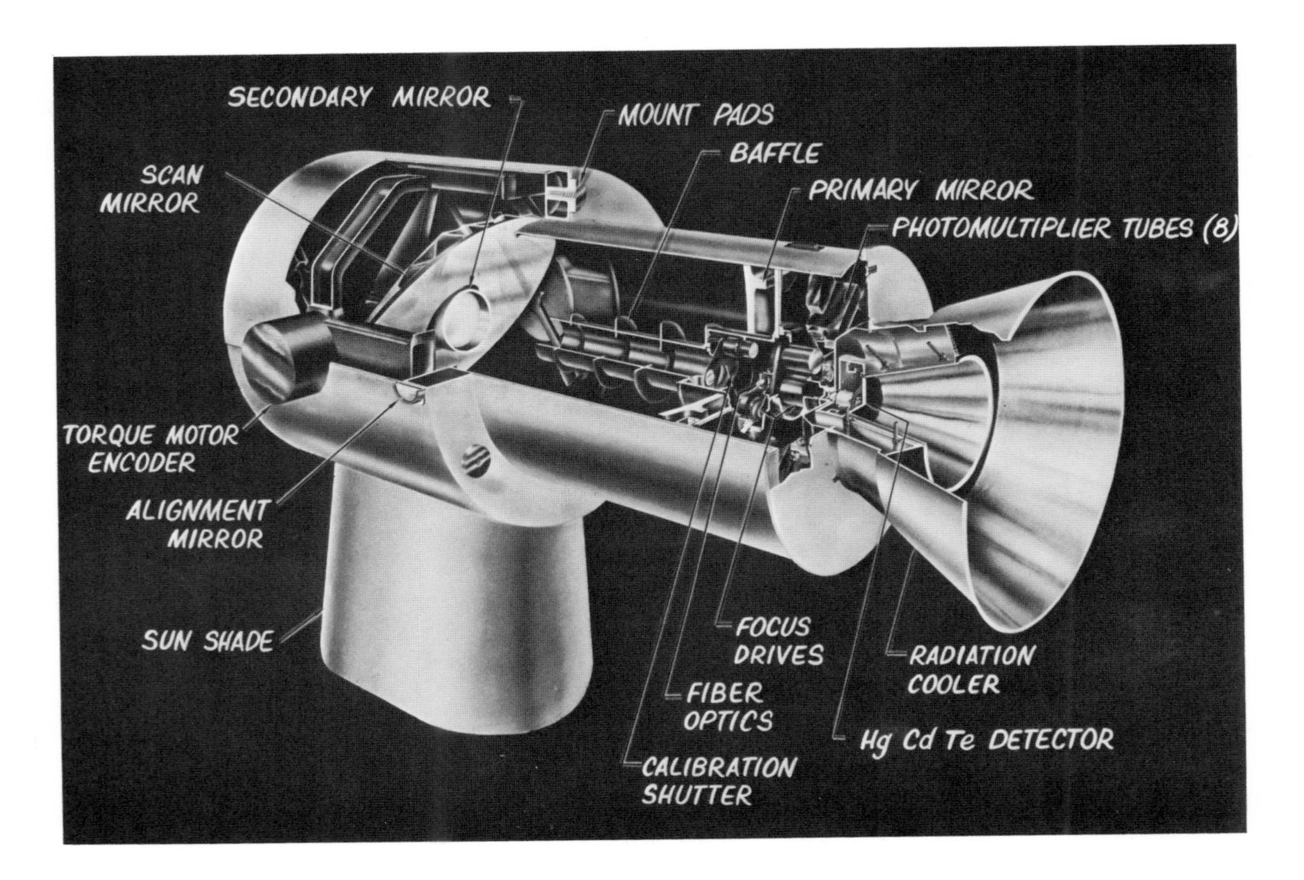

Fig. 5.9. Cutaway view of visible/infrared spin-scan radiometer for use in synchronous meteorological satellite.

Fig. 5.10. The earth as seen by SMS.

ible part of the spectrum and two cooled mercury–cadmium–telluride (Hg–Cd–Te) detectors, one of which is redundant, in the infrared. To cover the entire disc of the earth requires a little more than 18 minutes. The resulting daytime images consist of 14 568 lines which corresponds to a resolution of approximately 0·9 kilometres; the infrared images have 1821 lines and approximately 7 kilometres resolution.

The scanning data is transmitted to a Central Data Acquisition Station on the ground where it is processed in various ways and then transmitted back through the satellite to the National Environmental Satellite Service Headquarters in Suitland, Maryland. The central weather facility can have the processed signals forming a complete image only a second after the spacecraft has finished acquiring it. Fig. 5.10 shows the disc of the earth as seen by SMS. A band of cloudiness marks the inter-tropical convergence area and typical large-scale cyclonic activity can be seen over the North Atlantic. Fig. 5.11 is a sector covering the Atlantic Ocean east of Cuba and the southern United States—frequently a breeding area for hurricanes. By making a film strip of successive images it is possible to watch the movement and development of individual clouds, as well as large weather features.

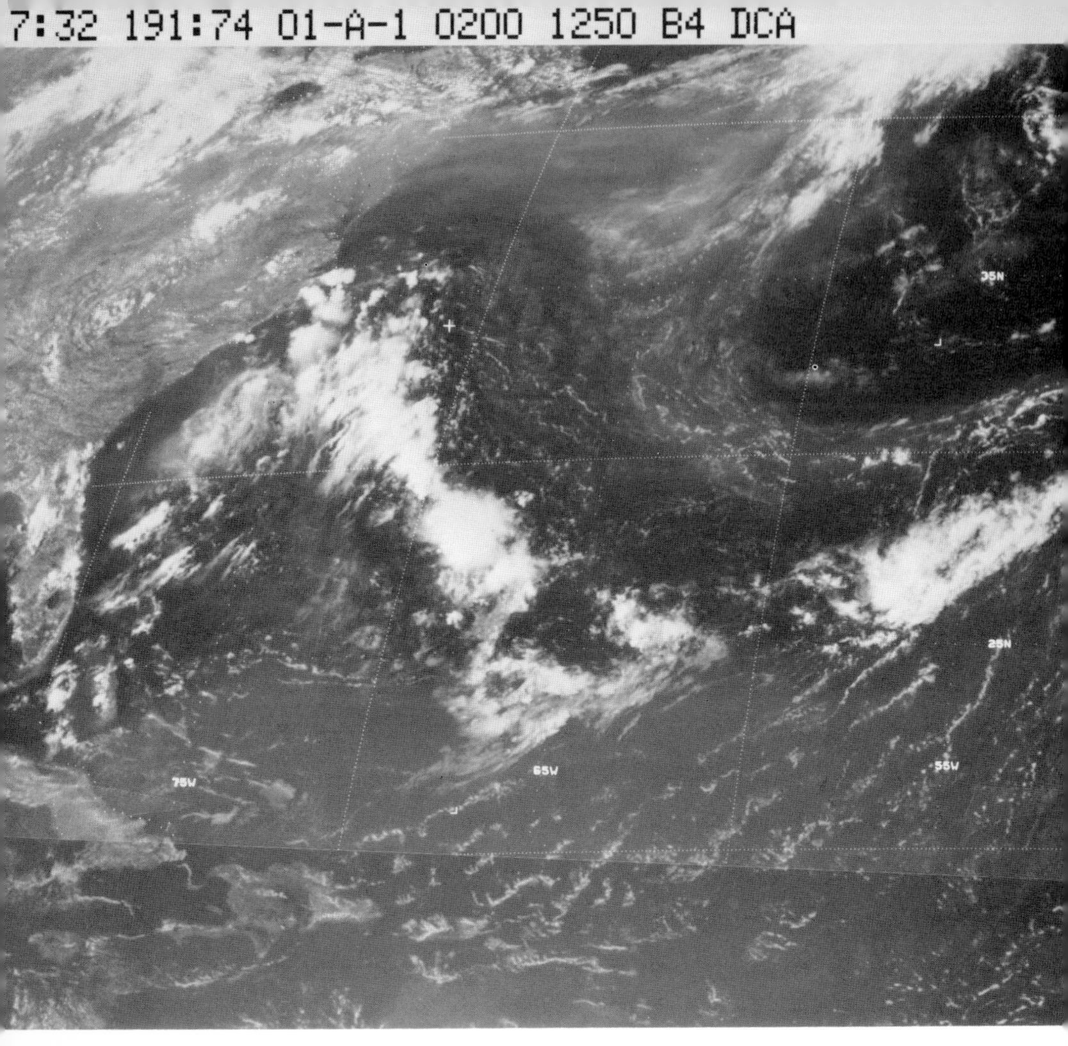

Fig. 5.11. SMS image of Atlantic Ocean east of Florida and Cuba. Breeding area for hurricanes.

Microwave techniques

In as much as all clouds except thin cirrus are opaque to both infrared and visible radiation, images made at these wavelengths show only the cloud tops and cannot distinguish between clouds that contain heavy precipitation and those that do not. For this reason, an imaging device has been developed using radio waves (at a wavelength of 1·55 cm) that can penetrate most clouds but not those that contain large water droplets, snow, or hail. This device, the electrically scanning microwave radiometer (ESMR), has been flown on *Nimbus* and has produced good results. Directional receiving characteristics are achieved by a flat phased array antenna that forms a beam about 1·5° wide. The beam can be scanned perpendicular to the flight path, by an angle of 50° to each side of vertical. Ground resolution at nadir is about 30 kilometres.

The quantity which a radiometer measures is called the 'brightness temperature', which is the product of the physical temperature and the emissivity. Thus neither of these quantities can be determined independently unless the other is known. However, in most scenes, the emissivity variations contribute more to the change in the

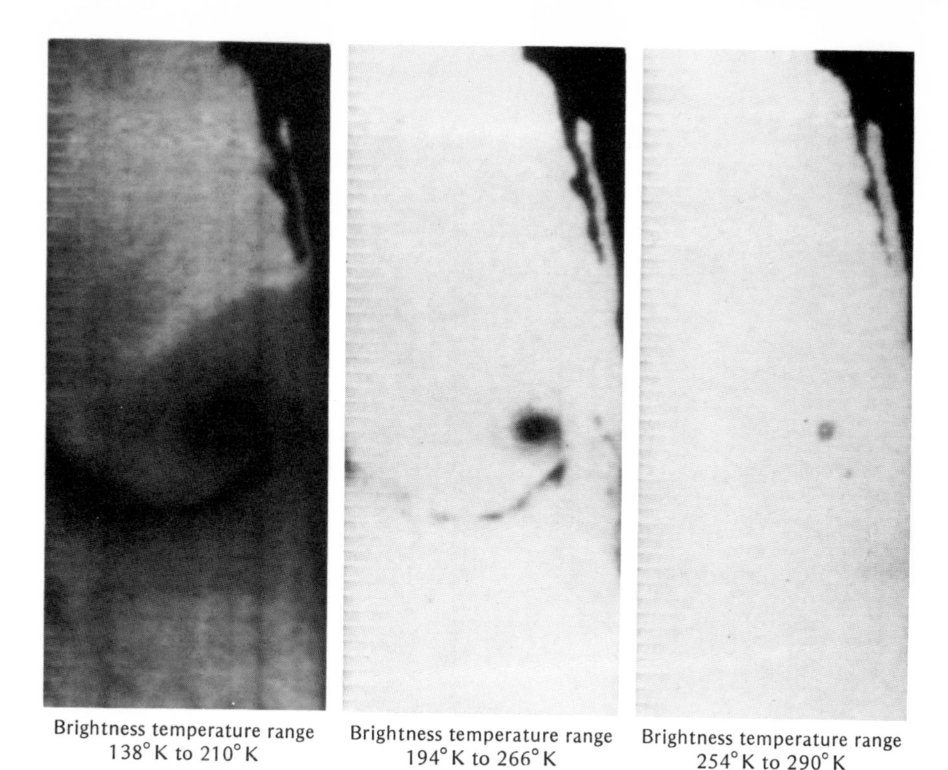

Fig. 5.12. ESMR images of Hurricane Ava south of Baja California, 7 June 1973, *Nimbus 5* orbit 2396. Left-hand image shows atmospheric moisture and light rainfall. Centre shows moderate rainfall. Right-hand image shows only heavy rainfall. Dark area at upper right in all three images is land.

product than the temperature variations do, at least in the microwave part of the spectrum. Therefore land areas, which have high emissivity, are generally dark. Open water, which has low emissivity, is light. Precipitation, which is dark, is hard to see over land but easy to see over the oceans. By adjusting the photographic process so that only a part of the brightness temperature range to which the ESMR is responsive is spread over the entire grey scale, particular features can be made to stand out clearly.

Fig. 5.12 shows microwave radiometer images of Hurricane Ava (1973) processed in this way. The relationship of atmospheric moisture to rain to intense rain is evident. Fig. 5.13 compares an ESMR image in the 190–259 K brightness temperature interval with images from the two infrared channels, showing how precipitation areas are related to the cloudy areas and to atmospheric moisture content in a storm south of Saudi Arabia.

One other important use for ESMR is in mapping sea ice in the polar regions, which it can do even through ordinary non-precipitating clouds. Fig. 5.14 illustrates its use in observing the advance and recession of sea ice around Greenland during the late winter and spring of 1973. Sea ice, especially new sea ice, has rather high emissivity at this wavelength because of the pockets of brine that are frozen into it. Thus

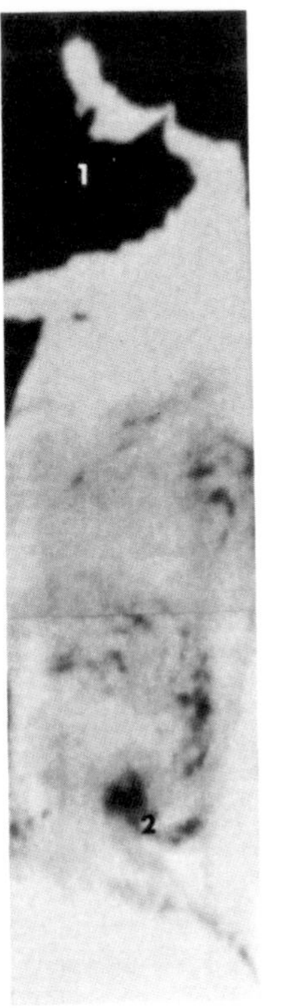

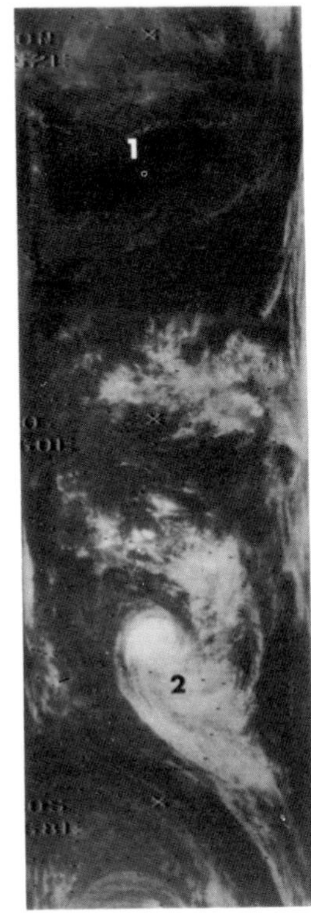

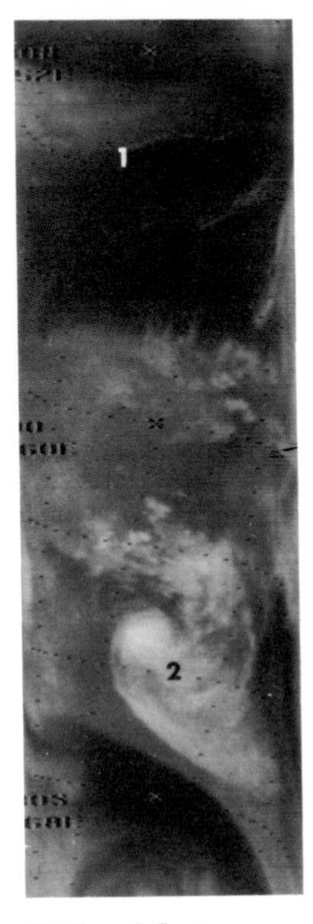

ESMR 1.55 cm

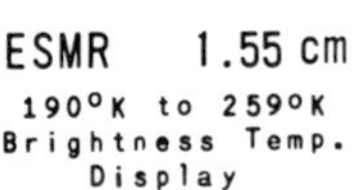

THIR 11.5 µm

THIR 6.7 µm

Fig. 5.13. Comparison of ESMR 190–259 K image (left) with THIR images for 'window' channel (centre) and water-vapour channel (right). In all three images, high brightness temperature is black and low brightness temperature is white. *Nimbus 5*, 20 December 1972, viewing storm (2) south of Saudi Arabia (1).

it appears dark like the land. Glaciers, on the other hand, have low emissivity; therefore the permanent ice cap, which covers much of the land area of Greenland, appears light in shade.

As this new microwave technology develops, resolution will be improved and active radar sensors will probably be used to determine the emissivity separately, making it possible to obtain both temperature and emissivity as quantitative measurements with high precision and sensitivity.

Determination of winds

The most useful input for numerical weather forecasting would be wind speed and direction measured at least once each day at about 500 kilometres intervals all over the earth at five to ten altitude levels. This data is not easy to collect; however, successive cloud images from a geostationary satellite can provide some wind data in areas where the right kind of clouds exist. This process can be automated using a man-machine system and has a typical accuracy of about 4·6 m/s for low-altitude

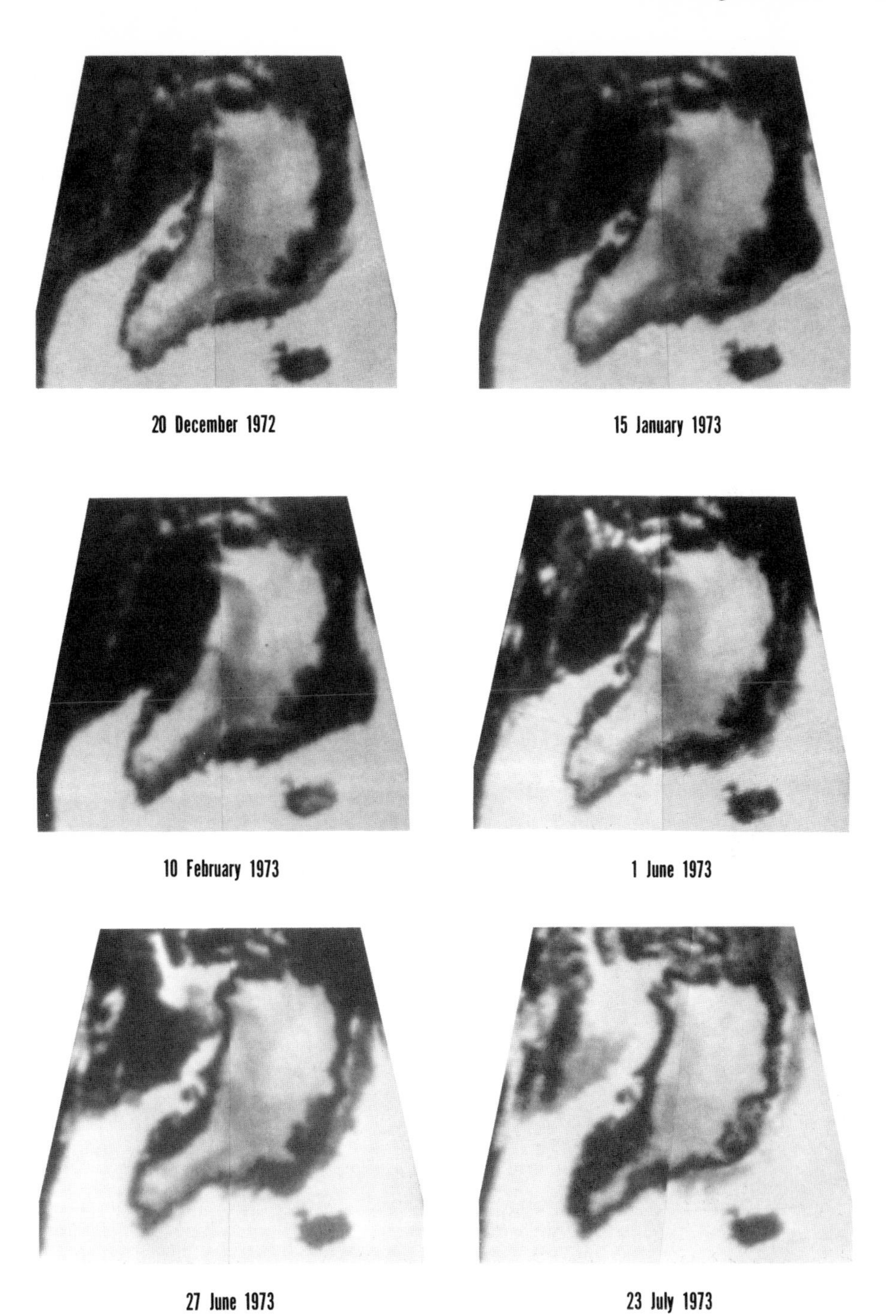

Fig. 5.14. ESMR images from *Nimbus 5* of Greenland ice cap and sea ice. New ice and snow appear black because they have high emissivity and therefore high brightness temperature. Open water appears white; old ice appears grey.

(1 kilometre) clouds, and 8·7 m/s for high-altitude (10 kilometre) clouds. There is usually some uncertainty about the height of the clouds used in this process, but this is being reduced as data from the SMS infrared channel begin to be used.

Balloons

An alternative method for remote observation of winds involves the use of tracer balloons that can be located by a satellite. The balloons used for this purpose are spherical, about 3·5 metres in diameter, and are made of very thin laminated sheets of a plastic, such as mylar. They are designed to float at a constant pressure altitude of 200 millibars† (approximately 3·7 kilometres) with an internal hydrogen pressure about 10 per cent greater than the outside atmospheric pressure. The average life of such a balloon is on the order of 100 days, not counting balloons of defective manufacture or those that are damaged during launching. Icing is the main cause of failure, especially over active storm areas.

At least four different methods have been devised to determine the position of these balloons from a satellite. Three of these involve radio measurement of the distance of the balloon from the satellite, the rate of change of this distance, or both. The fourth method makes use of the new *Omega* low-frequency worldwide navigation system. A simple receiver in the balloon picks up the *Omega* signals which are then transmitted to the satellite whenever it is nearby. These signals are converted in the satellite to balloon position and then transmitted to earth. Accuracy of location can be on the order of 1–10 kilometres for the various methods, which corresponds to errors in estimating average wind components of 0·15–1·5 m/s. It is possible also to make temperature, pressure, and humidity measurements by means of miniature instruments on the balloons, even to measure height above the surface using a 40-gram radio altimeter, and to transmit these data to the satellite along with the signals required for location.

More than 500 constant-pressure-altitude balloons have already been flown for test purposes, and hundreds more will be launched in connection with the first GARP (Global Atmospheric Research Programme) experiment. However, even if the balloons remain aloft for several months, the mean divergent flow of the atmosphere (away from the equator) at the 150-millibar level will generally remove them from the tropics, where they are most needed, in about 20 days. At a cost of $2000 to $10 000 (US) per balloon it does not now seem economically feasible to use tracer balloons on a continuing basis for routine weather forecasting. There is also some concern about the hazard these balloons and their payload may pose to aircraft.

A variant of the balloon system that has recently been tested successfully is the use of *Omega* dropsondes from stratospheric carrier balloons. This concept involves the use of large balloons capable of carrying 100-kilogram payload at a float altitude of 30 millibars (equivalent to 24 kilometres). This is well above icing levels and normal aircraft flight altitudes. Such balloons launched from a site near the equator are estimated to have a dwell time between 20° N and 20° S of more than 40 days. Each balloon would carry a dispenser with about 100 small dropsondes. On radio command from a geostationary satellite, a dropsonde would be released to descend by parachute at a rate of about 100 millibars in four minutes. Horizontal drift due

† A millibar is a measure of pressure used by meteorologists. It represents 1/1000 the standard atmospheric pressure at sea level.

to wind would be measured by the use of an *Omega* receiver on the sonde, the signals from which would be transmitted back through the carrier balloon to the geostationary satellite which gave the command. A pressure switch would give a signal at times when preset pressure levels were passed. In this way, vertical wind and pressure profiles over the tropics could be obtained more or less on command, especially in the areas most urgently needed. These big balloons with their payload of dropsondes would be relatively expensive, but not many of them would be required. They could reliably provide wind data at many levels of the troposphere but at only a few locations.

Temperature and humidity profiles

Although accurate closely spaced global wind measurements at several levels in the atmosphere would be the most important input for numerical weather forecasting, it does not seem possible to fully achieve this objective. Fortunately, except in the tropics, it is not necessary to have complete wind measurements if atmospheric temperature profiles are available. If the temperature is known as a function of altitude and if the pressure is known at some reference altitude, pressure and density profiles can be derived. From the horizontal variation of these pressure profiles and the hydrodynamic equations that are incorporated in the numerical model, the winds necessary to sustain the observed pressure variation can be obtained. (In the tropics, wind is only weakly coupled to the mass field and cannot be satisfactorily derived from the other parameters.)

Temperature measurements

Reasonably accurate temperature profiles can be obtained by infrared 'soundings' from satellites, with global coverage and spacing as close as desired except in areas of unbroken cloud cover. These soundings cannot all be made simultaneously; neither can wind measurements for that matter. However, with some additional complexity the computer models can be organized to accept input data for different locations at different times.

The temperature soundings are made by measuring the outgoing radiance in a vertical or near vertical column of atmosphere at five to seven different but closely

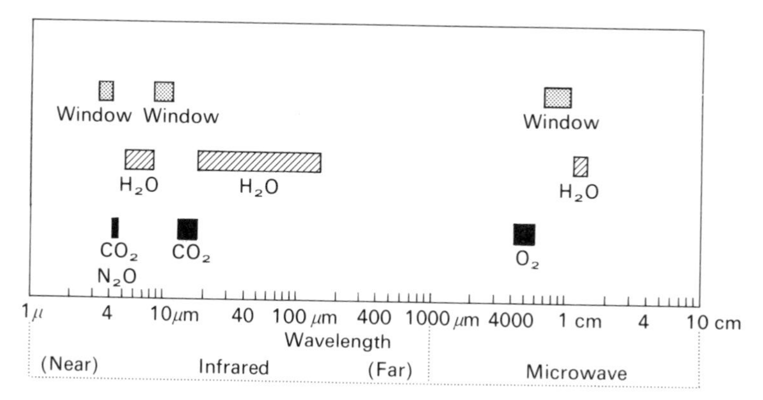

Fig. 5.15. Spectral regions found useful for indirect sensing from satellites. (GARP Publication Series No. 11 WMO-ICSU.)

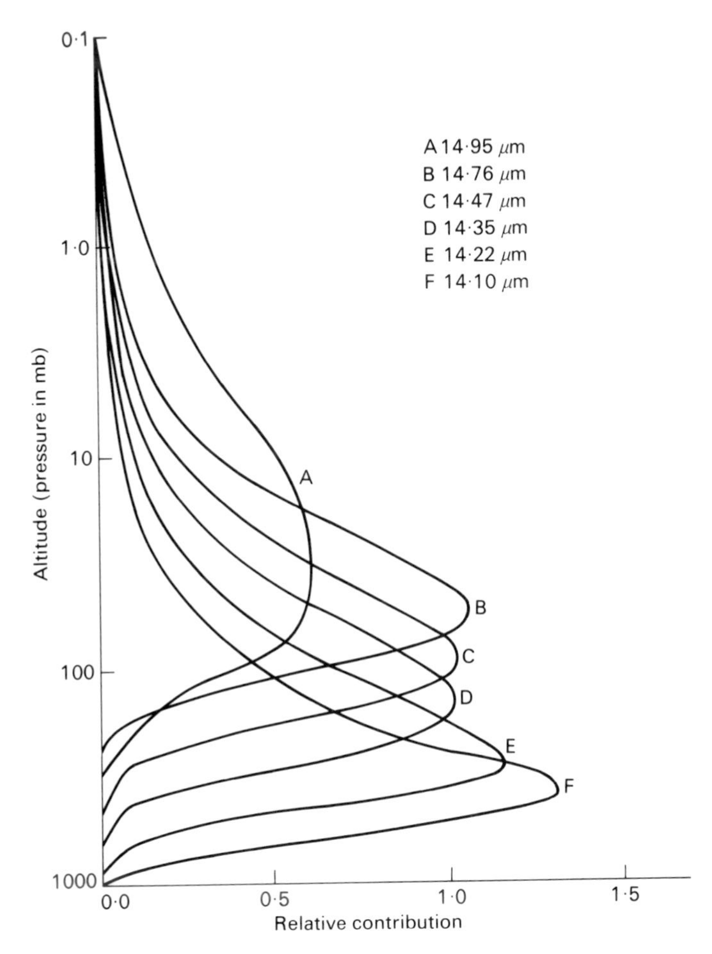

Fig. 5.16. Relative contributions of different levels of the atmosphere to outgoing radiation at six different wavelengths in the 15-micrometre carbon-dioxide band. (GARP Publication Series No. 11 WMO ICSU.)

spaced wavelengths in one of the absorption bands of carbon dioxide, plus one in a non-absorbing or 'window' region. As can be seen in Fig. 5.15, useful carbon dioxide bands can be found in the infrared at about 4·3 micrometre and 15 micrometre wavelengths. Window regions are at about 3·5 micrometres and 9 micrometres. At a wavelength near the centre of one of these absorption bands, the thermal radiation from any particular level of the atmosphere is more strongly absorbed by the carbon dioxide in the atmosphere above that level than at wavelengths near the edge of the band. Thus the radiance measured at the centre of the band will have originated high in the atmosphere and that at wavelengths successively farther away from the centre will have come from successively lower levels. Fig. 5.16 shows the relative contribution of different pressure levels to the total radiance at six different wavelengths in the 15-micrometre carbon-dioxide band. The radiation at any particular level is a strong function of temperature, and so a crude temperature profile could be obtained by assigning a temperature corresponding to the radiance measured at any wavelength

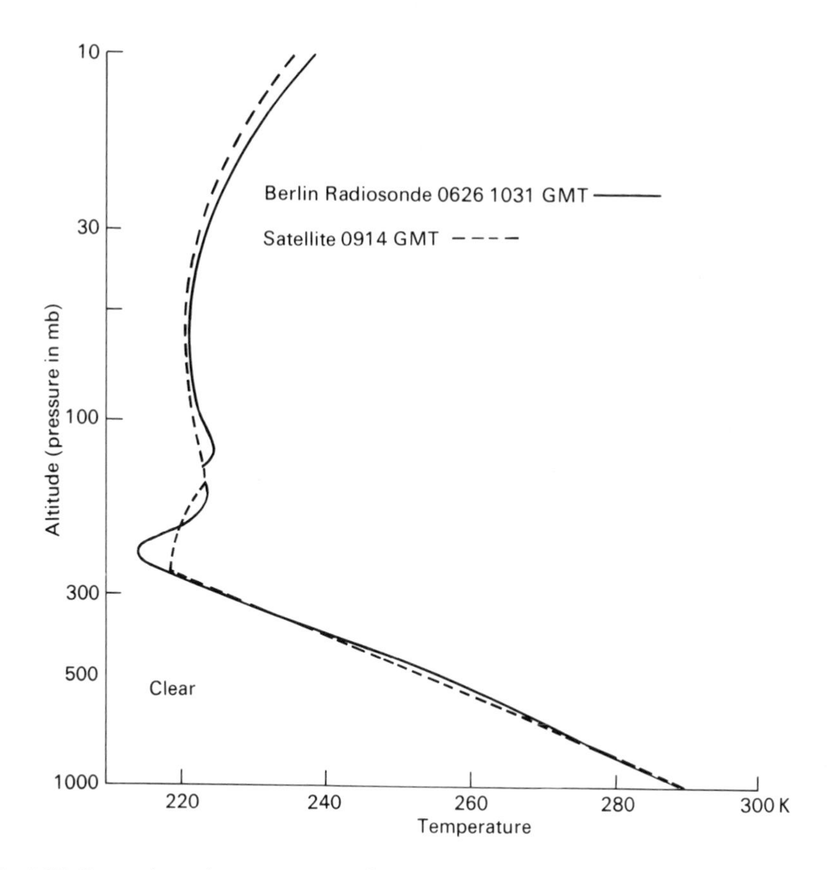

Fig. 5.17. Comparison of temperature profile calculated from infrared measurements over Berlin with profile measured directly by radiosonde. (GARP Publication Series No. 11 WMO-ICSU.)

to the pressure height of maximum relative contribution for that wavelength.

If the temperature profile were known, the radiance at each wavelength could be accurately calculated. By making an initial assumption as to the profile, comparing the calculated values for radiance with the measured values, and then adjusting the assumed temperature profile for best agreement, it is possible, in principle, to arrive at a reasonable approximation of the true temperature profile provided the profile does not have too complicated a structure and provided that the radiance measurements are accurate and not too 'contaminated' by clouds or other aerosols. Although it is not quite that simple, several different numerical methods for obtaining temperature profiles from radiance measurement have been developed and used successfully. The results, of which a typical example is shown in Fig. 5.17, indicate that the errors are generally within 1–3 °C, and for simple profiles less than 1 °C. Errors tend to be greatest near the surface, near the tropopause, or near any isothermal regions or inversions† that may be present. A window measurement is used to determine

† The tropopause is the boundary between the troposphere, the lowest region of the atmosphere, characterized by rapid vertical mixing, and the stratosphere, wherein vertical mixing is very slow. An isothermal region is one in which the temperature neither increases or decreases with altitude. An inversion is a region wherein the temperature tends to increase with altitude.

whether or not clouds are present in the field of view and to measure surface temperature in the clear areas.

The same process can be used in the microwave region of the spectrum using the oxygen absorption band near the 5-millimetre wavelength and the window at 7–9 millimetres. Carbon dioxide and oxygen are the two most useful absorbing gases to use for temperature measurement because their fractional abundance in the atmosphere is essentially constant up to at least 30 kilometres.

Water-vapour measurements

Water vapour, on the other hand, is present in the atmosphere in highly variable amounts. Its presence or absence can change the density of the atmosphere independent of temperature and pressure and its latent heat constitutes a major part of the energy transfer in the atmosphere; thus water vapour is in itself an important meteorological parameter the vertical distribution of which it is desirable to measure. Once the temperature profile is known this measurement can be made using the infrared and microwave water-vapour absorption bands shown in Fig. 5.15. As in the temperature profile measurements, the outgoing radiance is measured at several different wavelengths distributed between the centre and edge of the absorption band. The same integral equations apply; however, in this case, the temperature at each level is known and the set of integral equations is used to determine the vertical distribution of the water vapour. Taking into account probable errors in the temperature profile, the water-vapour profile can be determined with a vertical resolution of two or three layers and an error in average water-vapour content for each layer of 20–30 per cent or for total column water-vapour content of about 20 per cent.

The effect of clouds on vertical soundings of the atmosphere from satellites is not now considered to be as serious as it was when the technique was first proposed. Where clouds are continuous and unbroken, temperature profiles can at least be obtained down to the cloud tops. Analyses have shown, however, that by making a pattern of about 64 radiance measurements in each 400 kilometre by 400 kilometre grid square and comparing adjacent measurements to correct for partial cloud cover, temperature data with an accuracy comparable to or better than that of ordinary balloon radiosondes can be obtained all the way down to the surface for at least 90 per cent of all grid squares each day. Alternatively, a very-high-resolution slowly scanning radiometer can be used to find clear-column readings through holes in the cloud cover. Finally, microwave sounders can penetrate at least some clouds (cirrus, for example) and obtain temperature or humidity measurements at lower levels.

Typical meteorological satellites

Fig. 5.18 is a drawing of a recent *Nimbus* satellite, showing how the various instruments are installed and other design features. *Nimbus* is a three-axis stabilized satellite with the instrument ring always pointing downward. It is used for meteorological research and development of new types of instruments.

Fig. 5.19 is the ITOS-E satellite, one of the *Tiros* operational weather satellites. It is stabilized about two axes by the large momentum wheel shown at the top which spins at about 150 revolutions per minute. The spin axis is precessed by interaction of electric current in coils on the spacecraft with the earth's magnetic field, until it is normal to the plane of the orbit. Stabilization about the third or pitch axis

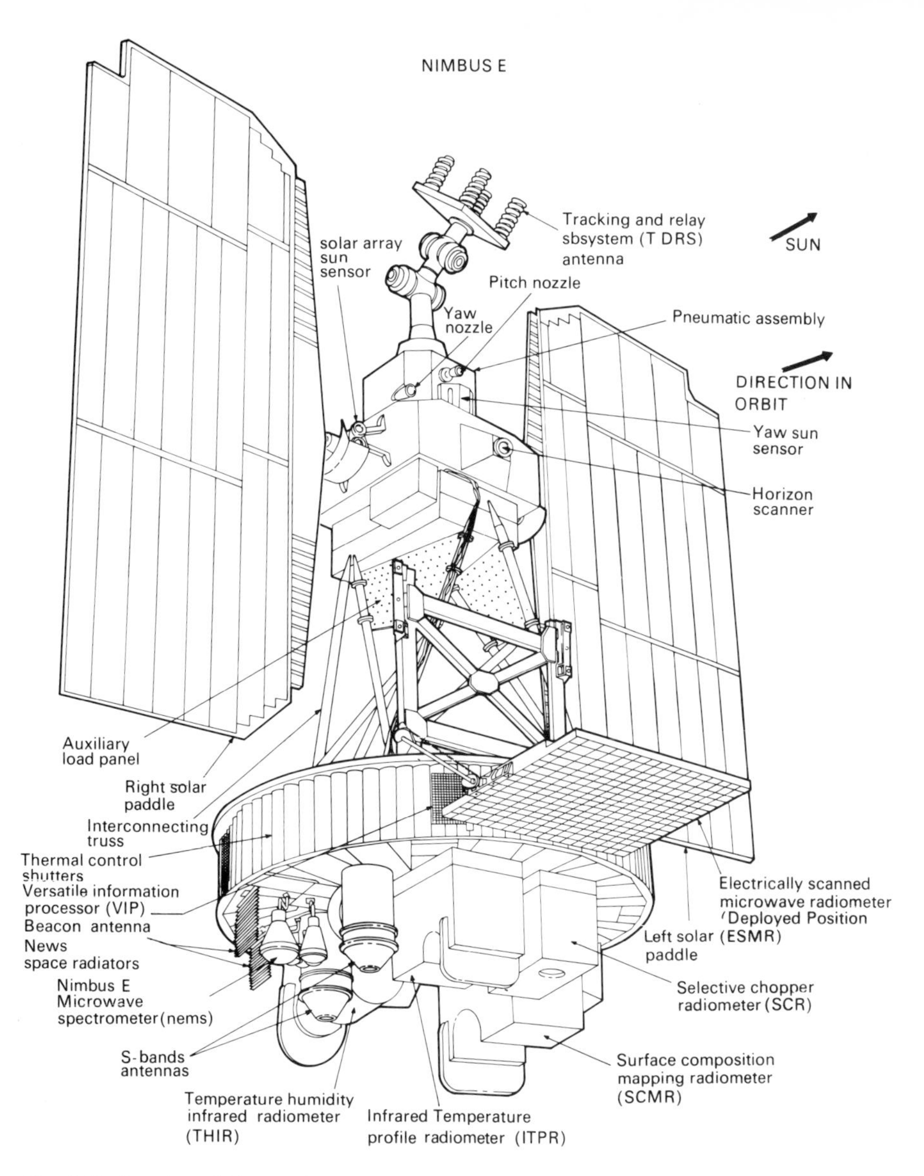

Fig. 5.18. Drawing of *Nimbus E* satellite showing installation of instruments.

is accomplished by applying a torque between the flywheel and the main body of the spacecraft until the latter is rotating at exactly one revolution per orbit and phased so that the surface on which the instruments are mounted looks straight down.

Fig. 5.20 is the synchronous meteorological satellite (SMS). The whole spacecraft spins at 100 revolutions per minute about an axis perpendicular to the plane of its orbit. The large oblong opening is for the spin-scan radiometer VISSR.

Mention should also be made of the Soviet Union's *Meteor* satellite series. *Meteor* is an operational weather satellite, based on at least six previous experimental meteor-

Fig. 5.19. Photograph of ITOS E satellite.

Fig. 5.20. Artist's conception of synchronous meteorological satellite, SMS.

Fig. 5.21. Meteor satellite at USSR Exhibition of Economic Achievement, Moscow, 1967.

ological satellites in the *Kosmos* series. It is a three-axis stabilized cylindrical spacecraft with sun-oriented solar panels on each side, as shown in Fig. 5.21. Although details of its payload of instruments have not been released by the Soviet Union, it is believed that it contains television cameras and infrared scanners for obtaining cloud images as well as infrared multichannel radiometers for temperature profiles. A total of ten *Meteor* launchings occurred between March 1969 and December 1971, and the programme seems to be continuing. These satellites have been credited with lives saved as a result of storm warnings, improved irrigation through better snow-cover data, and a 10 per cent reduction in sailing time for all Soviet shipping by making it possible to chart favourable courses with respect to storms, sea states, winds, and ice conditions. The developers of the *Meteor* satellites were awarded the Lenin Prize in science and technology for 1970.

Long-term forecasting and the climate

Despite all the jokes we make about the weatherman and his blunders we do depend on his daily forecasts and find them correct more often than wrong. Satellites are already contributing extensively to the accuracy of forecasts and to the length of time over which they can be made. What we should like most, however, would be the ability to predict months to years ahead whether the climate in a particular region will be unusually hot or cold, wet or dry, windy or calm, whether the growing season will be long or short, whether there will be enough snow in the mountains for a good skiing season or not enough. Timely and dependable answers to questions such as these would be of enormous economic value. Until now, we have had no better source for this kind of information than the *Farmer's almanac*. Furthermore, we are concerned about the very long-range effects on global climate of human activities such as burning fossil fuels, flying aircraft in the stratosphere, deforesting large areas of land, and creating large artificial lakes or reservoirs. We should like to know whether climatic changes such as the current devastating drought in the Sahel region of Africa is only a temporary dry spell or a permanent trend and, if so, what could be done to reverse it. Eventually, governments may want to attempt large-scale modification of the earth's climate—for example, by removing the permanent ice cover from the Arctic Ocean. We shall certainly want to be confident of our ability to forecast reliably the results of any such activity before it is undertaken.

A first step towards a better understanding of the climate and its changes is the making of models, based on physical principles, that show how various parameters interact to cause climatic change. The simplest of these models are the 'global-average' models, which use average values of the climatic variables over the whole earth and assume radiative–convective equilibrium. More complex models have been constructed involving the general circulation of the atmosphere, interaction with the oceans, and the effect of land masses, ice, and snow, and including non-linear feedback effects. Even the simplest models, however, have provided theoretical estimates of the effect of atmospheric carbon dioxide and aerosol content on the global average temperature and other basic relationships.

Radiation from the earth

It is generally agreed that radiation processes are among the most important determinants of climate and its changes. If the earth and space are to be in equilibrium, the total radiation input must equal the reflected radiation, which is mostly in the visible region of the spectrum, plus the thermal radiation emitted by the earth and its atmosphere into space. Because of the analogy to family income and expense this relationship is often called the radiation budget equation. If it does not balance, the earth must be gaining or losing energy and therefore getting hotter or colder, or storing energy or removing it from storage in the form of latent heat in the polar ice caps. Therefore accurate long-term measurements of the elements of the earth's radiation budget are of prime concern to the climatologist. These can be obtained directly by instruments mounted on satellites, although not without some difficulty.

In order to determine the solar constant to a high accuracy over a long period of time it would be desirable to use a primary standard. No such instrument suitable for the space environment has yet been designed, but there seems to be no compelling reason why it could not be. In addition to total solar energy incident on the earth, it would also be desirable to measure separately the energy in four or five contiguous spectral bands, extending from about 0·16 micrometres to about 4·0 micrometres. Calibration could be improved by recovering and recalibrating the instruments after a period in orbit. Direct measurement of the reflected and emitted radiation should be made simultaneously with scanning, narrow-beam radiometers and with accurately calibrated wide-angle (flux-density) radiometers. Reflected radiation should be measured in about six bands of the 0·16–3·5 micrometre region and the emitted radiation in two bands covering the interval from 4·0 micrometres to 50·0 micrometres. The long-term accuracy must be approximately 0·1 per cent. A combination of near-earth polar orbiting and geostationary satellites will be necessary to provide adequate sampling.

Cloud observations are important in order to obtain quantitative information about their statistical occurrence and distribution in relation to the important circulation parameters that are used in the models and to learn more about how their optical and radiative properties vary with cloud type, thickness, and structure. Clouds not only reflect sunlight back out into space, but also absorb, emit, and scatter radiation, thereby playing a crucial role in the energy budget. Fortunately observations of clouds as well as of snow and ice cover and other phenomena which change the surface reflectivity are already being made by research and operational meteorological satellite systems. Aerosol particles—by which is meant liquid droplets, ice crystals, smoke, dirt, and other solid particles thinly distributed through the atmosphere rather than concentrated in well-defined volumes like clouds—are much more difficult to measure. An instrument which measures the intensity and polarization of scattered solar radiation at a number of different wavelengths and scattering angles has been flown experimentally on an aircraft. It could be redesigned as a spacecraft instrument to provide the optical information about aerosols needed for radiation-balance studies and also to characterize the aerosols as to average droplet size, shape, and distribution. For vertical profiles of particle concentration in the lower troposphere, however, a pulsed laser system is needed. It is understood that the Soviet Union has tested such a system on a satellite.

Minor gaseous constituents of the atmosphere which play a significant role in the study of climatic changes include ozone, which efficiently absorbs solar ultraviolet radiation giving rise to the high-temperature region in the upper stratosphere, and carbon dioxide and water vapour, which, along with ozone, absorb and radiate strongly in the infrared at lower levels. The vertical distribution of these gases can be measured by the method previously described for measuring water-vapour profiles and also, in the stratosphere, by measuring the absorption of sunlight or starlight passing through the atmosphere near the horizon.

Although the variables which affect the radiative energy budget are clearly of primary importance there are some other observations which can be of interest. The change in the volume of ice stored in glaciers and the polar ice caps is both an indicator of changing climate and a source of energy and water storage. Changes in the average rainfall, in the extent or character of vegetative ground cover and in ocean currents and temperatures may all be useful. Furthermore, there seems to be evidence from growth rings on trees that the climate is correlated in some way with the sunspot cycle, and there are also other correlations which suggest an influence of solar–terrestrial physics on weather and climate. Although there is no accepted physical explanation for these effects it seems likely that they will be the subject of further attention as our understanding of the atmosphere increases.

Measurement of climatic variables must be essentially global in coverage and continuous over long periods of time. There would be little hope of obtaining such data by any other means than instrumented earth satellites. Although the serious study of climatology has only just begun, it is perhaps not too early to suggest that the greatest economic and social value of the space programme may ultimately result from its contribution to a better understanding of 'why the weather alters so'.

6 Helping the navigator

The winds and waves are always on the side of the ablest navigators.
Edward Gibbon *Decline and fall of the Roman empire,* Chapter 68

Since the time of Henry the Navigator some of the finest fruits of science and technology have been employed to improve man's ability to steer his ships, and more recently his aircraft, from one port to another with safety and accuracy. The magnetic compass, spherical trigonometry, the first star maps, the chronometer, the sextant, all were developed originally as aids to navigation. With the advent of electronics came radio beacons, radar, sonic depth finders, and accurately synchronized pulsed radio systems such as *Shoran, Loran, Omega,* and many others. Development of nearly frictionless bearings made possible the gyrocompass and precision inertial navigation equipment. It would seem that the modern navigator, with his radar, *Loran* receiver, and inertial navigation system, might already have almost more help than he needs! However, satellite systems, which can combine position fixing at almost any desired level of accuracy with communication capability, seem to offer an attractive alternative to these older systems.

The United States Navy navigation satellite system

The first and currently the only operational navigation satellite system is the *Transit* system developed prior to 1964 by the Applied Physics Laboratory of Johns Hopkins University for the United States Navy, which has a rather obvious interest in high-accuracy navigation. It uses a 62-kilogram satellite in a high inclination, nearly circular orbit at about 1100 kilometres. This satellite is stabilized in an earth-pointing attitude by a passive gravity-gradient technique (see Chapter 3) and the antenna pattern is shaped so as to provide a reasonably uniform signal level over a large surface area. Radio transmissions at 150 megahertz and 400 megahertz are controlled by a high-precision crystal-controlled oscillator on board the satellite. Ground installations track the satellite, compute its orbit with high accuracy, and transmit a series of future satellite positions to the satellite, where this information is stored in a small magnetic core memory. The satellite then transmits its position every two minutes by modulating both the 150 megahertz and 400 megahertz transmissions.

A ship can find its own position by using these satellite positions together with precise measurement of the frequency of the radio signal received from the satellite. This frequency is changed slightly by the relative motion between the satellite and the ship, and the change, which is called Doppler shift, is directly proportional to the rate of change of distance between these two points. The same principle applies to sound waves. Car drivers know that the sound of a horn from an approaching vehicle has a higher pitch than if both vehicles were standing still, and that the pitch suddenly decreases as the vehicles pass. This familiar experience also verifies the fact that the change in pitch (frequency) is proportional to the relative speed of the two vehicles. By keeping track of the Doppler shift in the transit satellite signal as received at the ship over the precisely two-minute interval between position transmissions, the navigator, or his electronics equipment, can determine the net change

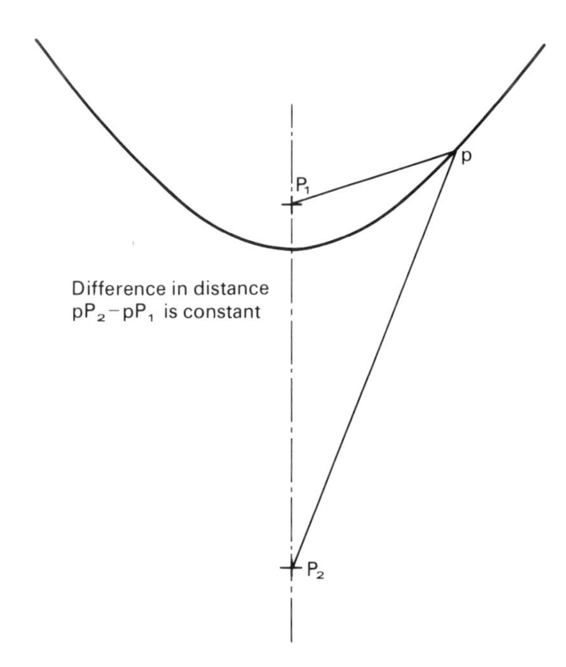

Fig. 6.1. Construction of an hyperbola.

in distance between ship and satellite corresponding to the two given satellite positions.

If we draw a curve on a piece of paper, as in Fig. 6.1, so that for every point p on the curve the difference in the distance to two fixed points P_1 and P_2 is constant, the curve turns out to be one that is called an hyperbola. In three dimensions it would be a surface generated by rotating this hyperbola about an axis passing through the two points. The positions transmitted by the satellite provide the two fixed points and the difference in distance from the ship to these two points can be obtained using the Doppler frequency shift. Thus the ship must be on a specific line of position (LOP) determined by the intersection of the hyperboloid of revolution with the earth's surface. This is LOP-1 in Fig. 6.2, obtained from data recorded while the satellite was going from P_1 to P_2. Using the data for a second interval, P_2–P_3, a second line of position LOP-2 can be generated. The intersection of these two lines of position is a fix. Since a satellite at this altitude travels roughly 900 kilometres in two minutes, an adequate baseline for the hyperbolic position-finding system is available from just these two two-minute observation intervals on a single satellite. If it is significant, the ship's own motion can be included in the solution.

The *Transit* system is able to compensate for two major potential sources of error. One is the slow drift in frequency of either the transmitter in the satellite or the receiver on the ship. This is eliminated by treating the frequency drift as a third unknown, the first two being the two position coordinates. Use of a third two-minute measuring period makes it possible to solve for all three unknowns simultaneously. The other correction is for ionospheric propagation effects. The propagation effects vary with radio frequency in a known way, and so an accurate correction for them can be determined by comparison of the Doppler shift at 150 megahertz and 400

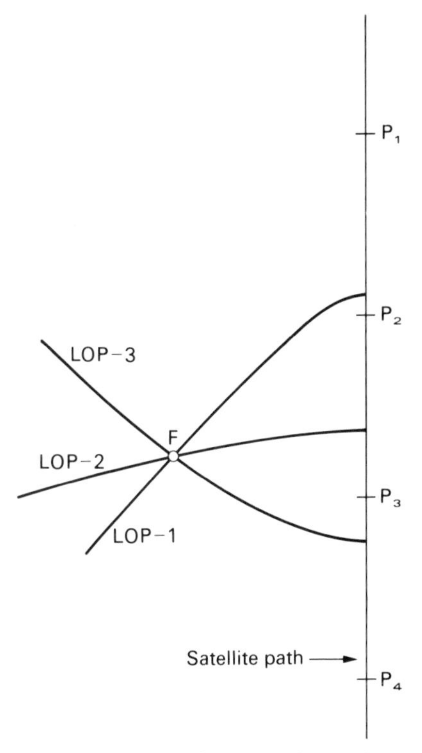

Fig. 6.2. Navigation fix using hyperbolic lines of position derived from three successive *Transit* satellite intervals.

megahertz. By making these and a few other less significant corrections, a global accuracy of about 0·1 nautical mile (185 metres) has been achieved.

This system provides very accurate navigational fixes for any number of users simultaneously from a single low-altitude satellite, which is obviously an advantage. However, fixes can be obtained only when the satellite is within range of the user, that is, about 3000–4000 kilometres. This means that with one satellite only two or at most three fixes at roughly 100-minute intervals can be obtained during each 12-hour period by a user at sea level near the equator. In order to fill in the gaps, four satellites in appropriately spaced orbits are used for the complete system. This makes it possible for a user anywhere on earth to obtain a fix not less frequently than about every 100 minutes. Should any of the satellites fail the overall system would continue to be useful but with twice daily gaps of several hours duration in availability. Engineers would say that the system degrades 'gracefully'.

Navigation systems based on geostationary satellites

Although systems like *Transit* that use low-altitude satellites can enable a vessel to determine its own position intermittently with great accuracy, they cannot do so continuously, nor can they easily provide communication capability or means to transmit the position information to a traffic-control centre or supervisory station on shore. Geostationary communication satellites can provide this communication capability. If at least two geostationary communication satellites are within range of a vessel they can be used to provide position information as well as communications. In principle, it would only be necessary for the vessel to send a signal to each of the satellites and measure the time for it to return, thereby obtaining the distance from the vessel to each of the satellites. Knowing the positions and height of the satellites the vessel could then position itself at the intersection of two circles determined by the distances to the two satellites. In practice, however, the navigator on the vessel would probably not know the precise position of each of the satellites at the time of the fix. Geostationary satellites are not perfectly stationary: they do move a bit if the orbit has even a slight inclination or eccestricity. The navigator would also need to know the magnitude of ionospheric propagation errors, which vary diurnally as well as with season and solar activity. Furthermore, for purposes of traffic control and fleet management, it is sometimes important to have the position information known quickly and accurately at a control station, in reference to the position of other vessels.

A system which takes account of these needs has been tested, using VHF and L-band (UHF) transponders on ATS satellites 1, 3, and 5. In this system a central control station sends a time marker in the form of a short burst of audio-frequency tone followed by an address code as modulation on its transmitted signal. This is received and retransmitted without delay by the satellite like any communication signal. All of the vessels using this system, and a few fixed ground stations as well, are equipped with transponders which receive all transmissions from the satellite and are able to recognize their own individual address codes. The transponder addressed by the control station, after a precise delay, transmits the time marker back to the satellite which relays it to the control station. The control station receives both the satellite retransmission of the original time marker and the delayed transmission from the transponder on the vessel to be positioned. By subtracting the known transponder delay from the difference between the time of arrival of these two time markers, the control station can determine the distance of the vessel from the satellite and thus a circular line of position on the earth's surface or at any known altitude above it. Repeating the process with a second satellite gives another line of position and hence a fix. The whole ranging process requires only a fraction of a second and gives a measured accuracy of 0·1–0·24 microseconds, depending on the available signal-to-noise ratio. By making ranging measurement periodically on ground stations for which the position is known, corrections for errors caused by ionospheric propagation effects and inaccurate satellite orbital information can be determined.

This system has been operated for seven or eight months on the tanker ESSO Bahamas plying a route between Venezuela and various ports on the eastern coast of the United States. Only VHF was used for this test. A comparison of position fixes obtained by the satellite system with simultaneous radar and visual fixes while the ship was close to land showed an average error of 1·3 nautical miles, most of which appeared to be caused by variable propagation effects that could not be completely compensated. The position-fixing function required only a negligible fraction of channel time, and so the system was also able simultaneously to transmit good-quality digital voice, teletype, facsimile, and slow-scan television signals from the tanker, at various points on her route, direct to her home office in New York City. The system has also been used esperimentally on other ships, aircraft, and unattended buoys.

Most of this experience was at VHF frequencies because ATS 5, the only satellite which carried an L-band transponder, suffered a malfunction at orbit insertion which caused it to spin, at about 77 revolutions per minute, instead of pointing continuously toward earth. As a result of this spin the antenna beam pointed at earth for only about 50 milliseconds during each 780 millisecond rotation period. Nevertheless, by synchronizing the timing of the transmission of the time marker and adjusting the transponder delay so that the signals would arrive at the satellite during its brief operable period, it was possible to use this satellite to obtain data on L-band ranging. On the basis of the ATS 5 tests, it is estimated that a fully operational L-band system, for which ionispheric propagation errors would be much smaller than for a VHF system, would provide a navigational accuracy of about 0·1 nautical mile everywhere within range of the two satellites except near the equator. When the vessel is near the equator it cannot obtain an accurate position in latitude from any number of satellites in equatorial orbit, although its longitude determination is very accurate. For this situation, two more satellites in non-equatorial earth-synchronous orbits would be required.

A number of other systems using geostationary satellites have been proposed and worked on. These include the use of angle as well as range measurement from the satellite using interferometer antennae, hyperbolic position grids established by synchronized pulses from three or more satellites, and rotating narrow fan-beam techniques. Out of all this experimental work will undoubtedly come one or more operational systems, possibly one for commercial shipping and one for air traffic. Both will involve communication and traffic-control aspects as well as frequent determination of position. Expected benefits include closer spacing and therefore higher traffic density on the most economic air routes across the North Atlantic, reduction of crew and greater automation of commercial shipping, and the saving of lives through faster and more effective search-and-rescue operations following an accident.

7 Surveying the oceans and the land

I am Monarch of all I survey,
My right there is none to dispute.
William Cowper (1782).
Verses supposed to be written by Alexander Selkirk

If indeed the purpose in making surveys of the earth's surface were to achieve political dominion over all the areas surveyed, there would be many to dispute! Fortunately the purpose is not to achieve power but rather to increase man's knowledge and understanding of the surface of the earth, the ocean streams that girdle it, and the life that inhabits it, and to help governments conserve and manage the resources under their control most effectively for the benefit of their people. This chapter will be concerned with how satellite technology can be used to extend the capability of the geologist, the geographer–cartographer, the glaciologist, the hydrologist, and the oceanographer. A good place to start seems to be with a description of a satellite system that is beginning to be used for such purposes.

The earth resources technology satellite

There is certainly some truth in the old saying that one picture is worth a thousand words. As a matter of fact one reasonably clear snapshot does contain more information, in the theoretical sense, than many thousands of words. I have already described the use of cloud pictures in meteorology (Chapter 5). Although meteorological satellites were not primarily designed to make pictures of the surface of the earth in cloudless areas, they have been found to be useful in that role, despite the rather coarse resolution of their imagery. Also, as everyone must be aware, the early astronauts spent some of their time photographing the earth with high-quality cameras and brought back pictures of almost breathtaking clarity and brilliance. These soon aroused so much interest on the part of earth scientists of various kinds that it was decided to construct and put into orbit a satellite primarily intended for making pictures of the surface of the earth or, in the more elaborate language of officialdom, 'acquisition of repetitive multispectral images of the earth's surface'. This was the earth resources technology satellite, ERTS-1, recently renamed *Landsat.* It also had a second objective, namely, to relay data from remote automatic sensor stations at fixed locations on the ground.

If *Landsat,* shown in Fig. 7.1, looks very much like the weather satellite *Nimbus,* the resemblance is not coincidental. The *Landsat* configuration was derived from that of *Nimbus* and several of the same subsystems are used. It has a mass of 953 kilograms, an overall height of about 3 metres, and a maximum diameter of approximately 1·5 metres without the solar panels, which extend to a total span of almost 4 metres. A three-axis active attitude-control system maintains the alignment of the

Fig. 7.1. Earth resource technology satellite, ERTS-1, recently renamed *Landsat*.

spacecraft to within 0·7°, using the local vertical and the direction of motion as reference axes. Instantaneous angular rates are held to less than 0·04° per second. The attitude-control system also rotates the solar panels so that they are always approximately normal to the direction of the sun. These solar panels are covered with photovoltaic cells that can provide a total of about 500 watts electric power, part of which is used to charge batteries that keep the satellite energized while it is on the dark side of the earth. As will be mentioned later, it is necessary that the orbit be accurately maintained throughout the useful life of the satellite; therefore an orbit adjustment subsystem is provided. This subsystem consists essentially of three small rocket thrusters that can be activated by ground command when an orbit correction is required.

Although *Landsat* is similar to *Nimbus* in configuration, the sensors which make up its payload are quite different from those of *Nimbus.* Because of its intended mission, the *Landsat* payload consists primarily of two independent multispectral imaging devices. One uses television-like technology with vidicon tubes and is known as the return beam vidicon (RBV) camera subsystem. The other is a mechanical scanning device, referred to simply as the multispectral scanner or MSS. Both cover the visible part of the spectrum plus near infrared. In addition to these sensors, there are two tape recorders, wide-band telemetry transmitters, and a data-collection subsystem (DCS) that can receive and retransmit real-time data from up to 1000 unattended fixed ground stations.

The RBV camera subsystem contains three individual cameras that operate in different spectral bands, blue–green, yellow–red, and red–infrared (Bands 1, 2, and 3), determined by filters built into the lens assemblies. Fig. 7.2 shows the measured spectral response of these three cameras. The cameras are carefully aligned in the spacecraft to view the same 185 kilometre square ground scene.

Although image resolution is a difficult quantity both to define and to measure, the approximate ground resolution in a *Landsat* RBV image, under favourable conditions, is in the neighbourhood of 80 metres. Simple linear features, having good contrast, such as a road or highway, can be seen in the photographs even if their width is somewhat less than 80 metres; complex features, such as two parallel roads or streets, however, cannot be resolved unless the spacing between them is greater than this distance.

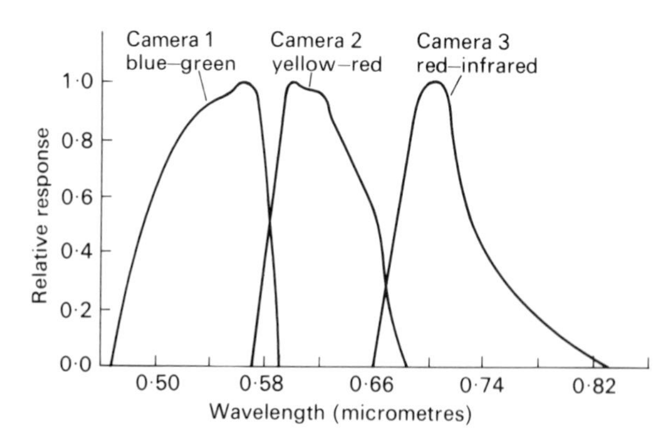

Fig. 7.2. Spectral response to the return beam vidicon (RBV) camera subsystem on *Landsat.*

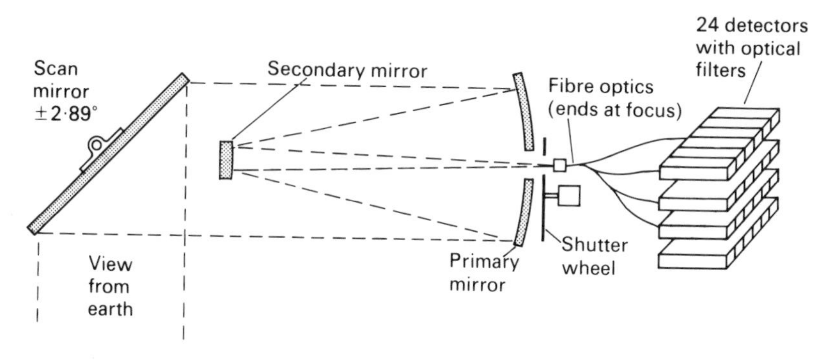

Fig. 7.3. Diagrammatic representation of *Landsat* multispectral scanner subsystem (MSS).

The other imaging device, MSS, is shown schematically in Fig. 7.3. A reflective optical system creates a real image in full colour, including infrared, at the focal plane near the rear of the instrument. As the scan mirror oscillates through an angle of 2·89°, the portion of the earth's surface being imaged sweeps over a distance of 185 kilometres in a direction perpendicular to the motion of the satellite.

At the centre of this image is placed a 4 X 6 array of optical fibre ends, each having a square aperture of a size that corresponds to 79 metres by 79 metres at the earth's surface. The six-fibre dimension of the array is positioned perpendicular to the direction of scanning. Each fibre terminates in a separate optical filter and detector, resulting in 24 simultaneous video outputs. One row of six optical elements is sensitive to yellowish–green colours (Band 4). The second is sensitive in the yellow–red region (Band 5), the third in the deep–red region (Band 6), and the fourth in the near infrared (Band 7). Spectral sensitivity in each of these bands is shown in Fig. 7.4.

The six video outputs from each row of six fibres thus represent six parallel scan lines across the scene, perpendicular to the motion of the satellite, and there are four

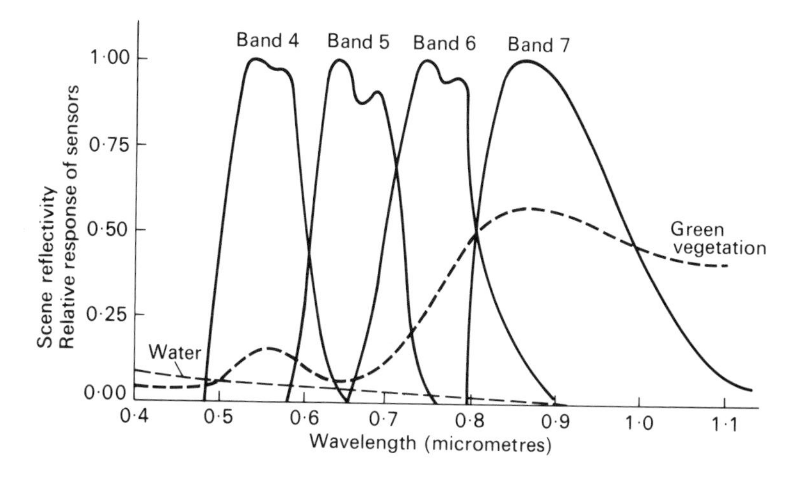

Fig. 7.4. Spectral response of multispectral scanner subsystem (MSS) on *Landsat* compared with reflectivity of vegetation and water.

of these six-line scans each in a different spectral band or colour, one following right behind another. By proper selection of the scanning time, the spacing between the four rows of fibres, and the precise time at which each video signal is sampled, each six-line scanning strip is made exactly contiguous to the one before, and the signal in each of the four spectral bands is exactly in register (positioned correctly) for each of the more than 5 000 000 picture elements (pixels) in a 185 milometre by 185 kilometre scene. A minor technological miracle, as any printer will confirm! Each sample of the MSS video data is converted into digital form and then combined into a serial data stream that can be transmitted immediately back to a ground station or stored on either of the two on-board tape recorders for transmission at a later time.

Video data from the RBV sensor subsystem is recorded and transmitted in its original analog form rather than being converted to digital.

The data collection subsystem (DCS) of *Landsat* acts only as a simple radio-relay station. It serves to collect data from small unmanned remote data-collection platforms (DCPs) that include up to eight sensors for measurements such as water level, flow in streams, water temperature, turbidity, relative motion (strain) of ice in a glacier, air temperature, humidity and wind velocity over a snow field, depth of snow, temperatures and temperature gradients in the cone of a volcano, concentration of specific pollutants in the atmosphere, or other quantities that it would be useful to monitor in relatively inaccessible locations. Data from the sensors is sampled periodically and transmitted about every three minutes as a 38-millisecond burst at 401·55 millihertz. If the satellite is within line of sight, it receives this transmission and retransmits it on the S-band telemetry link. If the satellite is also within range of a *Landsat* ground receiving site, the data is received, decoded, and transmitted along with other telemetry data to a central location where it is recorded on magnetic tape and disseminated to the users. Error coding is used to identify incomplete or invalid messages that may be caused by interference between two data-collection platforms transmitting simultaneously. Using three *Landsat* ground

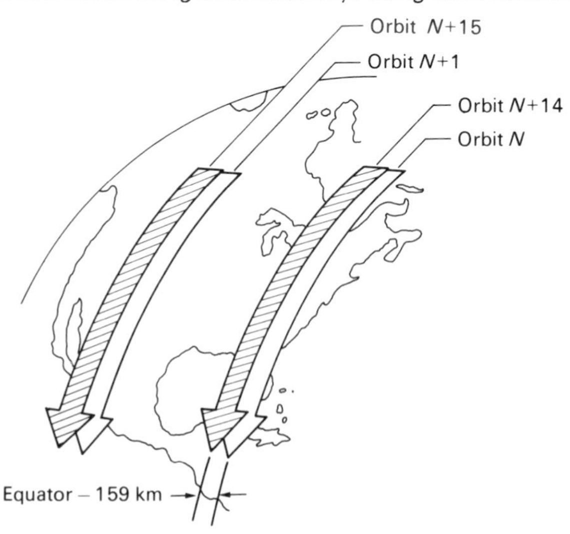

Fig. 7.5. Ground coverage pattern of *Landsat 1*.

receiving stations, one each on the east and west coasts of the United States and one in central Alaska, it is possible to achieve at least a 0·95 probability that one valid message will be received every 12 hours from any one of up to 1000 data-collection platforms located throughout north and central America.

The *Landsat 1* orbit is sun-synchronous, that is, it precesses at a rate such that it always passes over any point on the earth's surface at the same local or sun time (see p. 13). In this case, it always crosses the equator, heading south, at about 9.40 a.m. As the earth rotates with respect to the satellite orbit the path traced by the subsatellite point shifts westward by about 2869 kilometres at the equator during each orbital period. The period is precisely adjusted so that orbit 15 traces a path just 159 kilometres (at the equator) west of orbit 1 (see Fig. 7.5). After 18 days (250 orbits) the entire surface of the earth has been covered and the subsatellite point retraces the same path as on orbit 1. In other words, there is an opportunity to make an image of any particular 185 kilometre square on the earth's surface once every 18 days. Although the lighting conditions are always the same, except for seasonal effects at high latitudes, a specific area may be obscured by clouds when the satellite passes over and it may be necessary to wait one or more 18-day cycles until the opportunity will come again. There is, however, some overlap of the 185 kilometre strips. They shift westward only 159 kilometres at the equator, and so this overlap is 26 kilometres or 14 per cent at the equator, and increases to 44·8 per cent at the 50° latitude of central Europe. Thus there is a possibility to obtain an image of some small area on two successive days during each 18-day cycle.

At a receiving station, data from the different bands for each imaging subsystem are separated and put in appropriate form for further processing. They are then recorded on tape and the tapes are physically transported to a special *Landsat* data-handling facility. There the signals are corrected for known geometric and radiometric distortions and transformed into photographic images. Selected images are further corrected and positioned more accurately with respect to geodetic coordinates by means of ground control points, the position of which is accurately known. When desired by the user, the corrected digital image data is recorded on tape in a form suitable for computer analysis.

Landsat 1 was launched 23 July 1972. During its first 18 months of operation, it was able to make usable images, with less than 30 per cent cloud cover, of most of the earth's land areas. As of 2 April 1974 more than 110 000 scenes had been acquired and approximately 1 200 000 photographic prints or transparencies and 25 000 computer-compatible tapes had been generated and sent to users.

Many interesting details can be obtained from a single black-and-white image of the signal from just one of the four spectral bands. Much more can be learned, however, from a point-by-point comparison of the images in two or more different bands especially between one in the visible and one in the infrared part of the spectrum. For example, Figs. 7.6 and 7.7 show south-eastern England during March 1973 in Band 5 (yellow–red) and Band 7 (infrared). Many features appear to be quite different. Deep, relatively clear water, such as the portion of the English Channel off the coast of Hastings in the lower right-hand corner, appears almost black in both images. The mouth of the River Thames, shown at centre right, however, appears relatively bright in the yellow–red band, but very dark in the infrared. This is because water of any depth or turbidity has low reflectivity in the infrared, but shallow or turbid water appears much brighter than clear water in the yellow–red wavelength band. Thus the bright areas in Fig. 7.6 north of the mouth of the Thames are shallow tidal

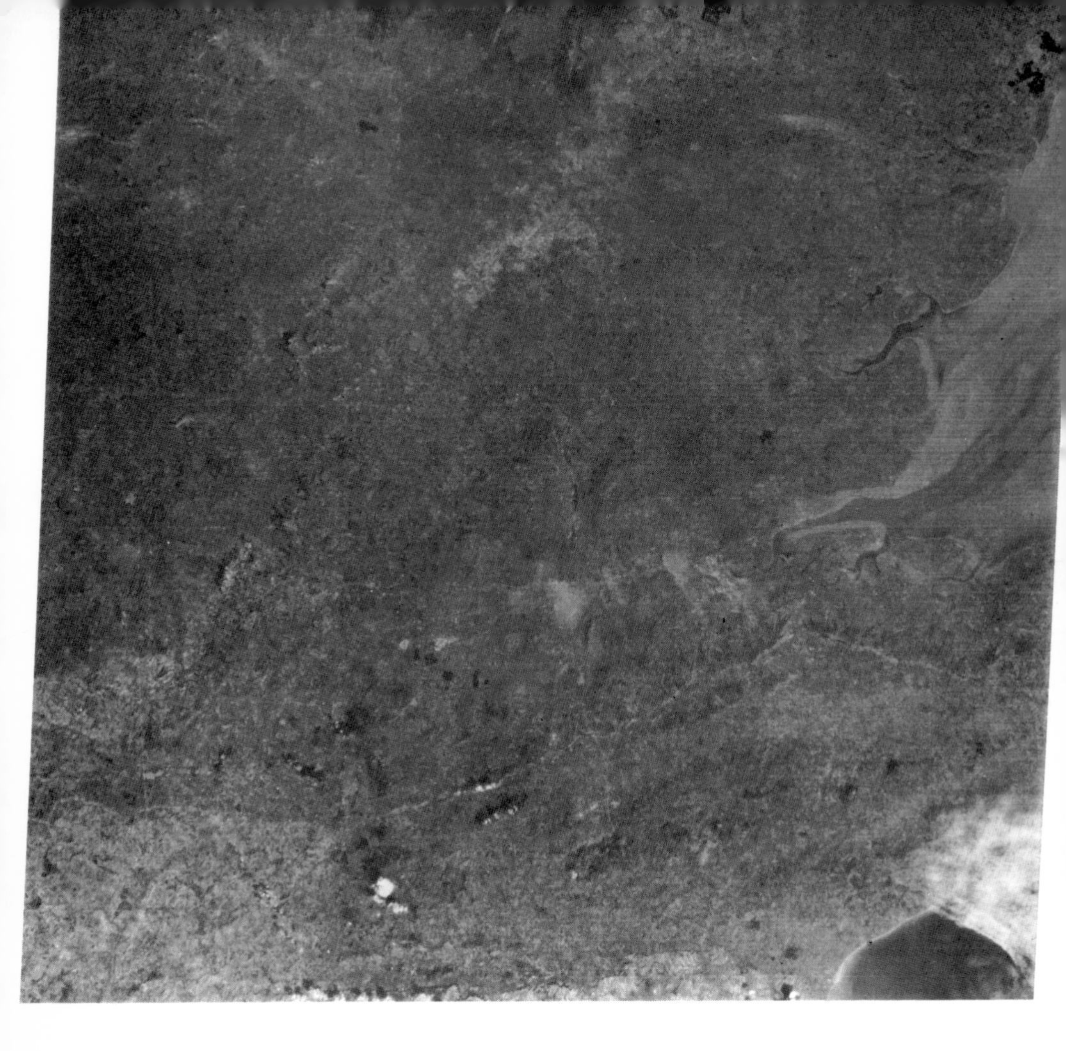

Fig. 7.6. Image of south-eastern England during March 1973 from multi-spectral scanner Band 5, yellow–red. (NASA ERTS E-1228-10293-5 01).

mud flats. The deeper channels cut by the Thames and by tidal flow, as well as the pattern of sedimentation and river effluents, can readily be seen. The heavily built-up city of London, almost in the centre of the picture, appears dark in the near infrared (Fig. 7.7) but is not so conspicuous in the yellow–red (Fig. 7.6), possibly because it is partially obscured by air pollution, several large sources of which are clearly visible. Streaks of smog from these sources can be seen streaming off to the south-east, carried by a north-westerly wind. For some reason, probably having to do with the particle size and composition, this smog is not very visible in the infrared picture.

Dry, relatively barren land, such as can be seen in the lower left-hand corner of this scene, is bright in both Fig. 7.6 and Fig. 7.7. Vigorous vegetation, which occurs higher up on the left-hand side, is darker in the visible band but light in the infrared. Areas which appear darker at the extreme top centre and upper right-hand corner in both images are forested with deciduous trees from which the leaves have fallen. The ground underneath is moist at this time of year and probably covered with old, dead leaves.

The different contrasts in the different spectral regions can be displayed simul-

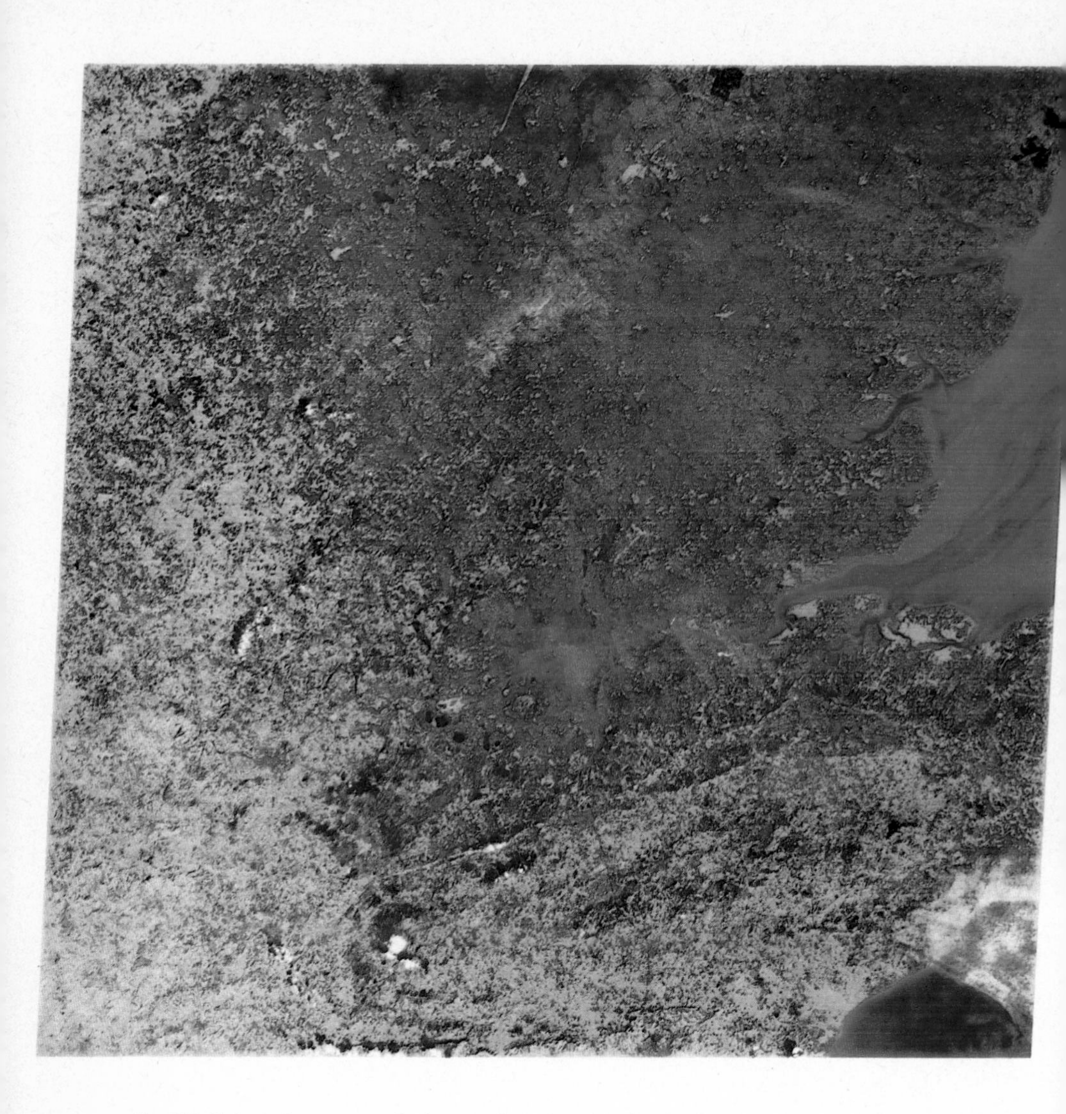

Fig. 7.8. False colour composite image of southeastern England during March 1973, from *Landsat* multispectral scanner. In this composite separate images from Band 4 (yellow-green), Band 5 (yellow-red), and Band 7 (infrared) are printed in blue (cyan), green, and red (magenta) respectively, superimposed on the same colour film. The reason for using a false colour combination such as this is to present in a recognizable form all the important information, including that from the infrared sensor which would be invisible to the human eye. London is almost at the centre. The mouth of the Thames River at center right shows channels cut by the Thames and by tidal flow in shallow mud flats and changing patterns of river effluent and sedimentation. (ERTS E-1228-10293-5)

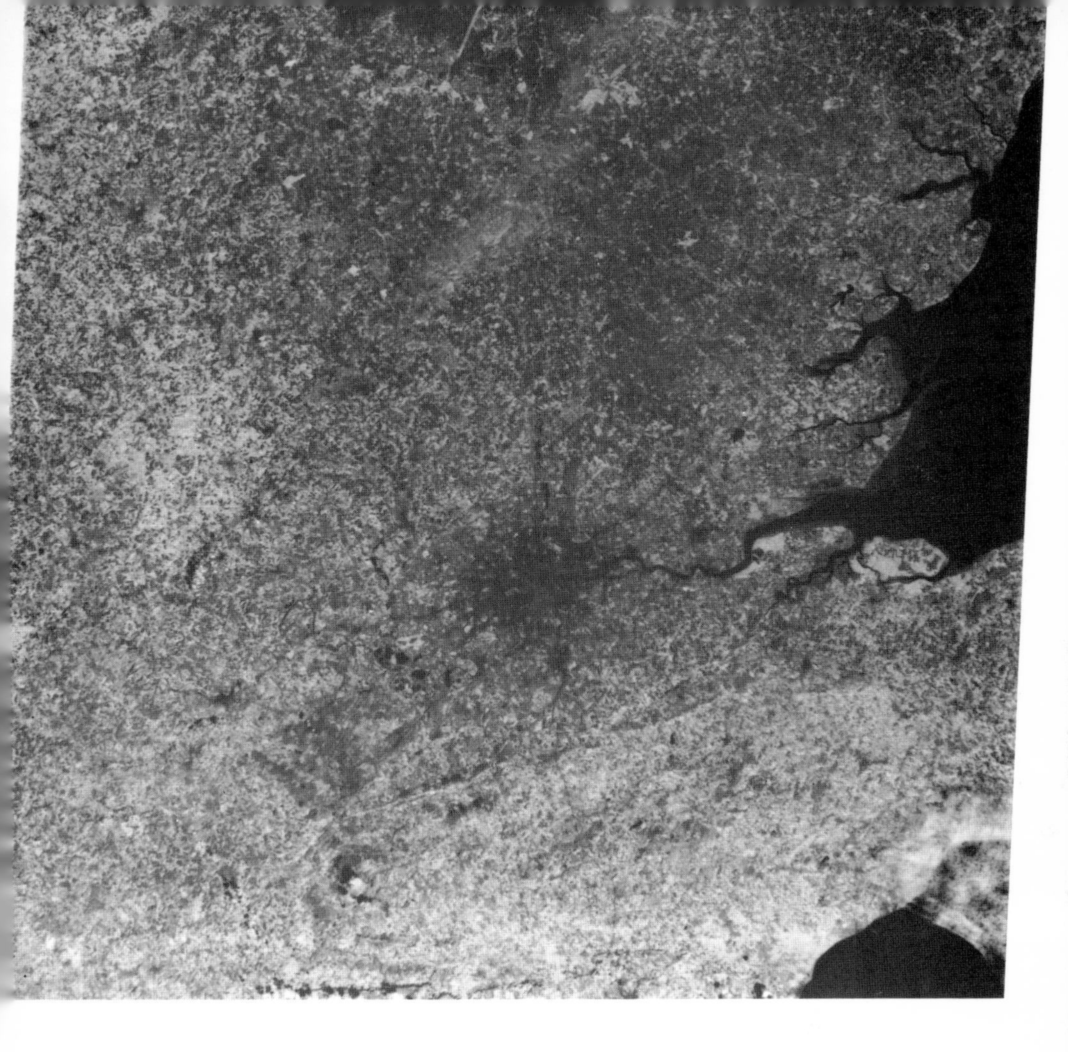

Fig. 7.7. Image of south-eastern England during March 1973 from multi-spectral scanner Band 7, near infrared. (NASA ERTS E-1228-10293-7 01.)

taneously in what is known as a 'false-colour' composite image. Here the yellow–green, red–yellow, and infrared bands are superimposed on colour film. The cyan (blue) dye in the film is used for the yellow–green (Band 4) image; green dye is used for the yellow–red (Band 5) image, which is not predominant in England at this time of year, and magenta (red) dye is used for the infrared (Band 7) image. Band 6 is not used. The resulting composite image for the same scene as in Figs. 7.6 and 7.7 is shown in Fig. 7.8. Note how much more clearly all of the features stand out. Another similar false-colour composite (Fig. 7.9) shows the Los Angeles Basin (lower right), the Mojave Desert (upper right), and snow-covered mountain ranges in southern California on 14 March 1974. The well-known San Andreas fault can be seen extending from left to right in the centre of the picture. Contrast between the metropolitan areas, the forested and grass-covered mountain slopes, the various alluvial materials covering the Mojave Desert, and the agricultural complex of the San Joaquin Valley (upper left) is very obvious.

There is so much detail in one of these images that it takes even an expert many hours just to begin to see what is there. The human eye, wonderful instrument that it is, cannot easily pick out many small similar details in a large scene. In any case, the sheer volume of useful information produced by this one satellite is enough to swamp the available human data-interpretation capability. Consequently attention has been given by several universities and industrial concerns to the development of automatic aids to data interpretation. One typical device, the General Electric Company's Image-100, uses the computer-compatible digital tapes produced by the ERTS data-handling facility to reconstruct an image on a monitor. An operator viewing the image can designate a particular point for which he may have ground truth or other information about the vegetation, soil, or water conditions, and he can then ask the machine to delineate all other areas having exactly the same spectral 'signature', and thus determine their total area. Working from the original digital data, the computer can do this quickly and with far better accuracy than any human eye. Other machines are being developed which will recognize desired geometric patterns, spatial frequency spectra, and so on. Only by the use of such machines can the full power of the earth-observation satellite system be applied to the broad range of human problems for which it was intended.

Making and correcting maps

A two-dimensional or planimetric map is the most basic information about an area that one can imagine. It shows the shape and location of natural features such as lakes, streams, shorelines, islands and mountain peaks, and also of cultural features, especially roads and cities. It also forms the basis for geological, agricultural, demographic, and other special-purpose maps. For other purposes, three-dimensional or topographic maps are required showing contours of elevation with respect to mean sea level or some other reference.

For many years, the data for such maps was collected by surveyors, hacking their way through forests and climbing mountains with chain, transit, and level. In this way they were able to locate principle features with reasonable accuracy and to fill in the remainder with a certain amount of imagination and inspiration, but it was a laborious, time-consuming, and sometimes dangerous process. During the last 35 or 40 years, aerial photography has largely taken over the surveying function. Special cameras have been developed, together with photo-interpretation aids that use stereoscopic viewing, accurate measuring machines, and electronic computers for making geometric and other corrections. Despite this advance, however, only about half the land area of the earth has been mapped adequately for most purposes. Generally speaking, the areas that are economically important have been mapped, although there is a continuing need for updating even in these areas. In the least affluent areas detailed mapping is almost non-existent. The underdeveloped regions will almost certainly remain underdeveloped until maps have been made, resources surveyed, and geographic studies completed.

Landsat images for cartography

Although the *Landsat* was not primarily designed for cartography, it is rather obvious that the images it produces can be used by map-makers, and that the images are especially valuable for inaccessible areas such as the Arctic and Antarctic regions and the jungles of South America and Africa. They are not very useful, however, for making topographic maps, because they do not normally produce overlapping images

Fig. 7.9. False colour composite image of southern California, U.S.A., during March 1974, from *Landsat* multispectral scanner. Colour substitutions used in this image are the same as in Fig. 7.8. The Los Angeles Basin is at lower right, the Mojave Desert upper right, and snow capped mountain ranges in the center. Contrast between the metropolitan areas, forested and grass-covered mountain slopes, alluvial materials covering the Mojave Desert, and the agricultural complex of the San joaquin Valley (upper left) is obvious. A trained eye can also discern the well known San Andreas fault system extending from left to right through the central part of the image. (ERTS E-1234-18021-5)

taken from different directions such as are needed for stereoscopic viewing. This is not an inherent limitation; the RBV subsystem could have been programmed to provide overlapping images in the along-track direction, but only at the cost of reduced coverage elsewhere. In any case, it was not done. The principal value of *Landsat* in cartography is in making large-scale planimetric maps and photo-image mosaics, and for making planimetric corrections to small-scale topographic maps.

For the preparation of maps in support of the United States Antarctic Research Programme, the United States has obtained aerial photographs over an area of about 3 250 000 square kilometres. This effort has required many years of work and cost many millions of dollars. About 100 frames of *Landsat* imagery now cover the same area, previously covered by more than 100 000 aerial photographs. However, only a small fraction of *Landsat* images have less than the 10 per cent cloud cover originally specified for cartographic purposes by the United States Geological Survey. Attempts are now being made to work with images containing up to 50 per cent cloud cover; however, it is clear that some regions will have to be covered several times by *Landsat* to obtain useable images. Also, it should be noted that the approximately 99° orbital inclination required for *Landsat* does not permit images to be made further north or south than 82° latitude, so a small area near the poles cannot be covered.

In the Antarctic areas that have been studied, significant map corrections have been found in the Drygalski Ice Tongue on the Victoria Land coast, in the size, shape, and position of the Thwaites Iceberg Tongue on the shore of the Amundsen Sea, and in the positions of Franklin Island and of the Ross Ice Shelf Front. New land features were also discovered in some regions. These studies have only just begun; however, they are being continued and expanded to cover all of the Antarctic continent that is available to *Landsat* imagery and also most of the Arctic region, including Alaska.

In another area of the world, *Landsat* images are being used to correct maps of the interior of Brazil, where significant errors were found in the location of some tributaries of the River Amazon. Some new streams and other previously unmapped features were found. Mapping based on *Landsat* imagery has been found useful in the planning and construction of the trans-Amazon highway, a major engineering and social reconstruction project in Brazil.

Intercontinental surveys

On a completely different scale, satellites are also useful to the map-maker because they can provide precise information about the shape of the geoid (see p. 100) and they can help in locating reference points precisely with respect to each other over distances too great for conventional surveying. Triangulation surveys on the earth's surface cannot be carried across seas of any appreciable width so, until the advent of satellites, surveys on any one continent were isolated from those on another. A further difficulty with conventional surveying has been that optical lines of sight close to the surface of the earth are bent by variable gradients of the atmospheric refractive index. Measurements of vertical distance have therefore been rather unreliable.

Included in the earliest United States satellite programme, during the International Geophysical Year, was a precision optical tracking system using a network of telescopic cameras of advanced design, known as Baker–Nunn telescopes. These telescopes were able to photograph even a small satellite against a star background using only reflected sunlight during periods just before sunrise or after sunset. With careful reduction of the photographs taken by these cameras, it was possible simultaneously to determine the orbits of the early satellites with great precision and also to locate the tracking stations with respect to each other to an accuracy that ultimately reached about 10 metres. Thus the first intercontinental ties were made between surveying systems in North and South America, Europe, Asia, and Africa. Also, from studies of the orbits of selected satellites, it was possible to improve greatly the existing knowledge of the shape and mass distribution of the earth. This work has been continued, using both optical and radio means of satellite tracking, one example of the latter being the United States Army SECOR satellite, which was specifically designed for surveying purposes.

The geoid

The geoid is an imaginary surface that could be thought of as the surface of the oceans and seas if they were completely undisturbed by tides, currents, or winds. Normally it is assumed to be an ellipsoid of revolution with a polar flattening of 1/298·24. Its exact shape is important, however, because it is the surface to which a spirit level is parallel and a plumb bob perpendicular, and its position with respect to the real surface of the earth is important because it is the reference against which all elevations are measured.

A long section of the border between Canada and the United States follows the 49th parallel (49° N latitude) and was astronomically determined. However, the 49th parallel is not a simple curve, as usually drawn on the maps, but instead is slightly sinuous because of geoidal undulations caused by the gravitational effect of mountains, valleys, and variations in the density of the earth's crust. These local gravitational anomalies can best be measured by making a fairly extended and dense network of surface measurements of the total gravitational field. From such measurements the local shape of the geoid can be determined. However, the grosser features, especially over the vast empty ocean areas, can better be determined by careful analysis of satellite tracking data. These two kinds of data, together with all available results from geometric satellite geodesy (precise locations of satellite tracking stations), are combined to produce the best overall data about the earth, which is updated periodically and is published as an international reference known as *The standard earth.*

Geodetic satellites

Several different kinds of geodetic satellites have been orbited. As early as 1962, the *Anna 1B* satellite launched by the United States Department of Defense incorporated a powerful, precisely timed strobe flasher, so that optical observations would not be limited to twilight hours, and so that precise crystal clocks would not be needed at the observation stations. It also had corner reflectors for passive radar tracking, a transponder for radio-range determination, and an ultra-stable radio beacon for Doppler-shift measurements. Best known, perhaps, are the NASA *Geos* series of satellites, most recent of which is the *Geos 3*, launched in April 1975. It is a relatively small gravity-gradient stabilized satellite that carried, in addition to radio tracking beacons, a special S-band transponder for making range and range-rate measurements from the ATS F geosynchronous satellite, a 13·9 gigahertz radar altimeter, and laser corner reflectors. By laser tracking, it is now possible to determine the position of the satellite and to make geometrical geodesy measurements with an accuracy of about 1 metre, in an area of reasonable size having four laser tracking stations at the corners. Such an area is defined by existing tracking stations at Greenbelt in Maryland, Bermuda, Grand Turk Island in the Bahamas, and Cape Kennedy in Florida. The laser-tracking solution is especially accurate for satellite altitude. Given an accurate value for satellite altitude, the radar altimeter is capable of measuring the height of the ocean surface to ± between 3 and 5 metres and can detect variations in height of about half a metre over a distance of a few hundred kilometres. This makes it possible to locate submarine trenches and sea mounts by their gravitational influence on the sea surface. Satellite-to-satellite tracking can be used to measure accelerations caused by local gravitational anomalies with an accuracy good enough to determine harmonic terms up to the 28th in the mathematical representation of the earth's gravitational field, that is, an anomaly of approximately a twenty-eighth of the earth's circumference, at the satellite altitude of 1100 kilometres. These latter two capabilities are new and will undoubtedly be improved in later satellite systems, such as *Sea Sat 1*.

Searching for mineral resources

Contrary to some overly optimistic beliefs, satellite instruments cannot 'find' metal ore deposits, coal, or petroleum. All they can do is to help the geologist survey a region from a different viewpoint and possibly enable him to identify some areas he might otherwise have missed where the probability of finding geological 'goodies' is relatively high. Usually, before any extensive drilling or prospecting is started, other confirmatory data such as gravitational, geomagnetic, or radioactive anomalies are sought and seismic tests are made, if appropriate.

What the geologist begins with is a photomosaic or photomap at a scale of 1:1 000 000 or even 1:5 000 000, and what he looks for first are lineaments (linear markings, not necessarily straight) that can represent faults, intrusions, or boundaries between areas having different kinds of rock or different tectonic (structural) history. He is interested also in surface fracture zones or subsidence areas that may indicate a history of tectonic activity below the surface. These geological and morphological features are not easy for the untrained eye to pick out, even in a good image. Of the various *Landsat* bands, the infrared Band 7 and the yellow–red Band 5 are most frequently used; however, in some situations geologists prefer Band 6, and in others the false-colour composite is easiest to work with. The main advantage of satellite imagery for this work is the large area that can be covered by a relatively few images with uniform tonality. Older geological features that may have considerable mineralogical significance are frequently covered or partly obliterated by sedimentation, sand, or volcanic debris. Only by looking at a small-scale image, that is, one that covers a large area, can these major features be separated from the many minor surface details that tend to obscure them in larger-scale images.

An example is the LaPaz mosaic (Fig. 7.10), compiled by the United States Geological Survey together with geologists from seven South American countries using *Landsat* Band 6 images made in 1972 and 1973. It covers an area that lies between latitudes 16° S and 20° S and longitudes 66° W and 72° W. South-eastern Peru lies in the upper left, northern Chile in the south centre, and south-western Bolivia comprises the eastern half. Major bodies of water shown are the Pacific Ocean in the south-west and part of Lake Titicaca at the north-centre margin; Lake Poopo

Fig. 7.10. La Paz mosaic, made from parts of about 20 ERTS Band 6 images covering the region 66° to 72° W and 16° to 20° S. (ERTS Experiment No. 189, NASA, Dept. of Interior, US Geological Survey.)

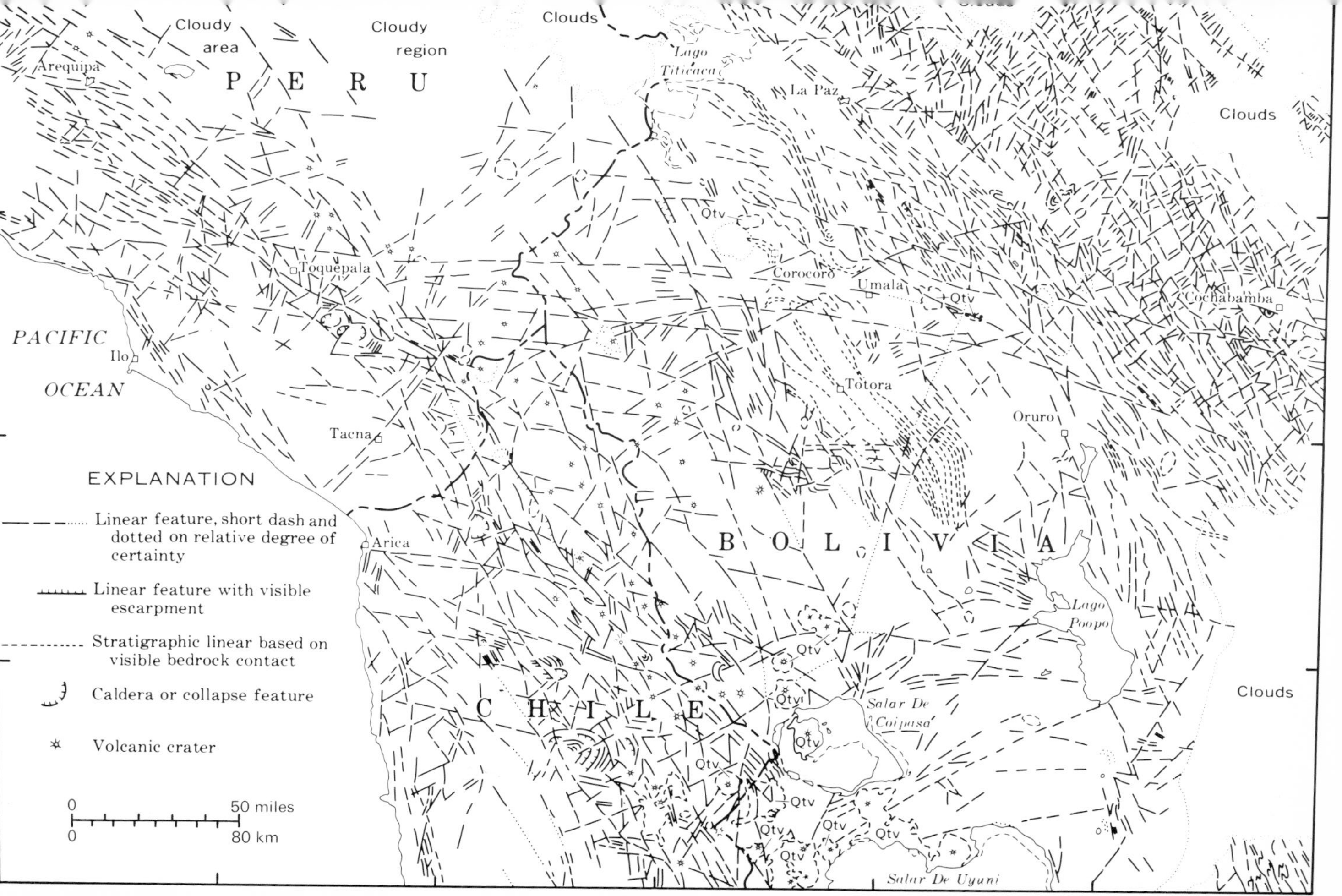

Fig. 7.11. Tectolinear map of the region shown in the La Paz mosaic, 66° to 72° W, 16° to 20° S.

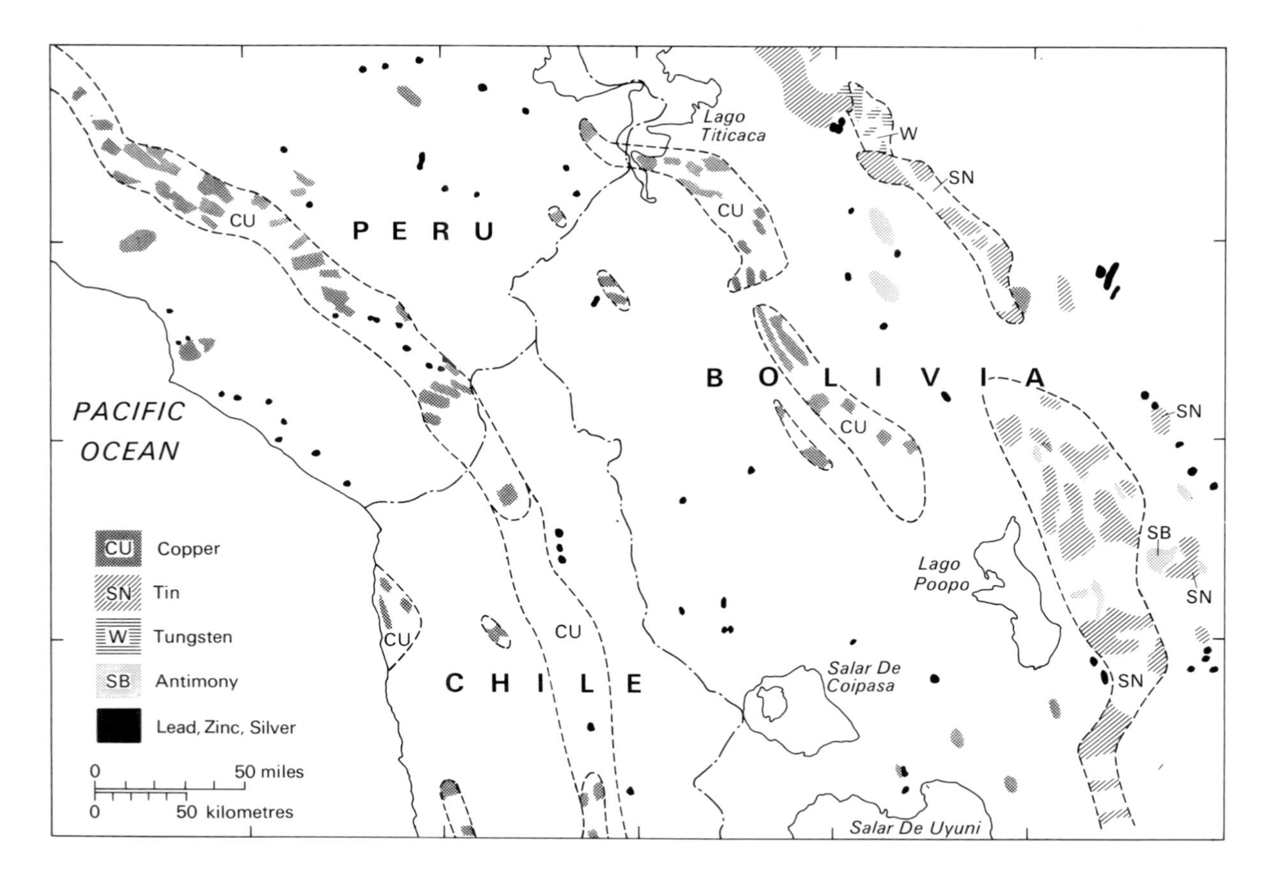

Fig. 7.12. Metellogenic map of area covered by La Paz mosaic, 66° to 72° W, 16° to 20° S.

and the Salars of Coipasa and Uyuni are in the south. The region is important for its mineral resources, for within it are major copper, tin, and tungsten deposits as well as significant production of lead, zinc, silver, gold, and sulphur. A map depicting linear and curvilinear features is shown in Fig. 7.11. Solid lines, long dashed lines, and dotted lines indicate relative degrees of certainty in interpretation. Short dashes indicate linear features believed to correspond to bedding plane contacts of significance. Volcanic centres are shown by a radial star-shaped pattern.

A composite metallogenic map showing the location and principal elements of known ore deposits, made from published maps and other information provided by Peru, Chile, and Bolivia, can be seen in Fig. 7.12. Comparison of this map with the tectonic overlay of Fig. 7.11 enables investigators to see the close relationship of the structural features to the location of ore deposits of various kinds. Most prominent are the close association of major copper deposits to structural linears in northern Chile (Mocha and Cerro Colorado) and in south-eastern Peru (Toquepala to Arequipa). This narrow band of apparently intensive structural activity seems to be a fertile area for further detailed studies. The LaPaz mosaic is only the first of twelve 4° X 6° areas in South America to be studied in this co-operative project.

Fresh water

Without fresh water, almost all land plants and animals, including man, would perish. Water, then, must rank among our most valuable earth resources, although we often take it for granted. Fresh water is found in rivers, streams, lakes, and underground aquifers which are replenished by rain and melting snow, or ice. Much of the potentially available fresh water is wasted, at least from an economic viewpoint, by evaporation or by running into the sea and mixing with salty water. Some of it is stored seasonally in mountain snow fields and glaciers; some of its is impounded in natural lakes and swamps and some in streams that have been dammed to create man-made reservoirs. Occasionally there may be too much water for a river system to carry away. When this happens, the result can be a disasterous flood. As the population of the earth increases, it becomes increasingly important to understand and monitor all the elements in this complex hydrological cycle. Timely information about available water resources has a direct and tangible benefit from the viewpoint of people who must plan and efficiently operate irrigation and flood-control projects, hydroelectric power systems, municipal water systems, river navigation, and sewage-disposal plants. It is also of value to planners concerned with the aesthetic and recreational needs of man and with ecological problems, not the least of which is the disappearance of marshes and wetlands.

Fortunately, water, ice, and snow are easy to delineate and to measure in satellite images. Remote rainfall, water-level, and stream-flow gauges can be placed at strategic locations, using the *Landsat* data-collection subsystem described earlier in this chapter to bring the data together at a place where it is needed. Also, as has already been described, new weather satellites have some capability to measure rainfall. Thus, for the first time, most of the required hydrological information is beginning to be available.

One obvious type of information that can be obtained from *Landsat* images is the number and size of small lakes and ponds in a region. Figs. 7.13 and 7.14 are images in Band 5 and Band 7, respectively, of a *Landsat* MSS scene containing several hundred of these features. With good viewing conditions, such as an enlarged image on a

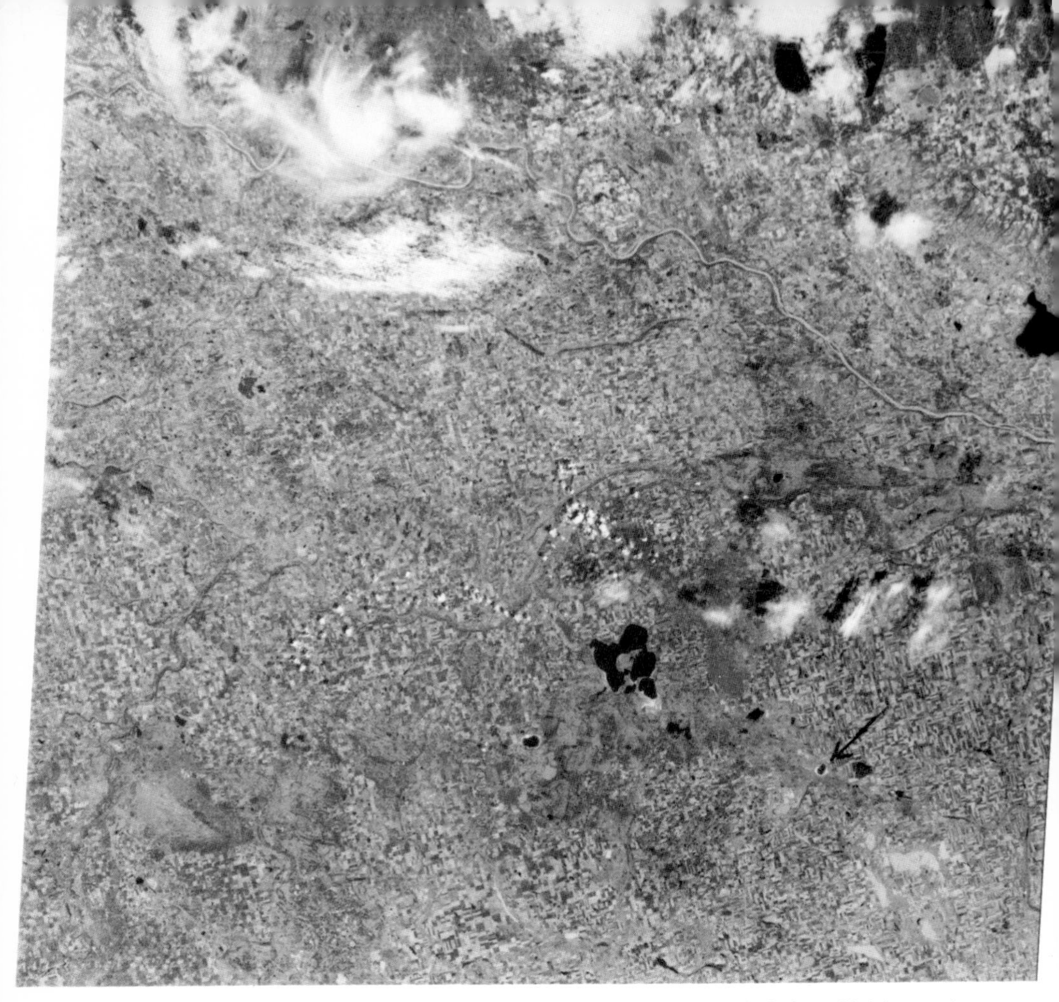

Fig. 7.13. Image of west-central Saskatchewan, Canada, 9 August 1972, from Multispectral scanner Band 5. (NASA ERTS E-1017-17494-5 01.)

light table, ponds as small as 100–150 metres across can be identified. In this reproduction, of course, only larger features can be seen. However, you will notice that a number of the lakes which are easily visible in Band 7 are almost indistinguishable in Band 5. These are the shallower ponds, in which the thin layer of water is almost transparent to the wavelength of Band 5 (yellow–red) but still very absorbent in the infrared of Band 7, and in which algae and water plants are most likely to grow during a hot summer (this image was made on 9 August 1972). Using computer-compatible image tapes, these lakes and ponds can be identified, counted, and classified as to depth; their surface area can be measured; and the volume of water they contain can be estimated. All this can be done automatically once the machine has been properly 'instructed' by its human operator. This data can be updated seasonally, if desired, a task which would be almost impossible by other means.

Groundwater, like mineral deposits, cannot be seen directly from satellites, but areas in which strong-flowing aquifers (underground streams of water) are likely to be found can usually be identified from the geology of the region. One geomorphic setting that can be correlated with major aquifers is the confluence of major old glacial meltwater channels. These settings are commonly associated with sand and gravel in large outwash and glacial-lake deltaic plains. The aquifer sites are usually

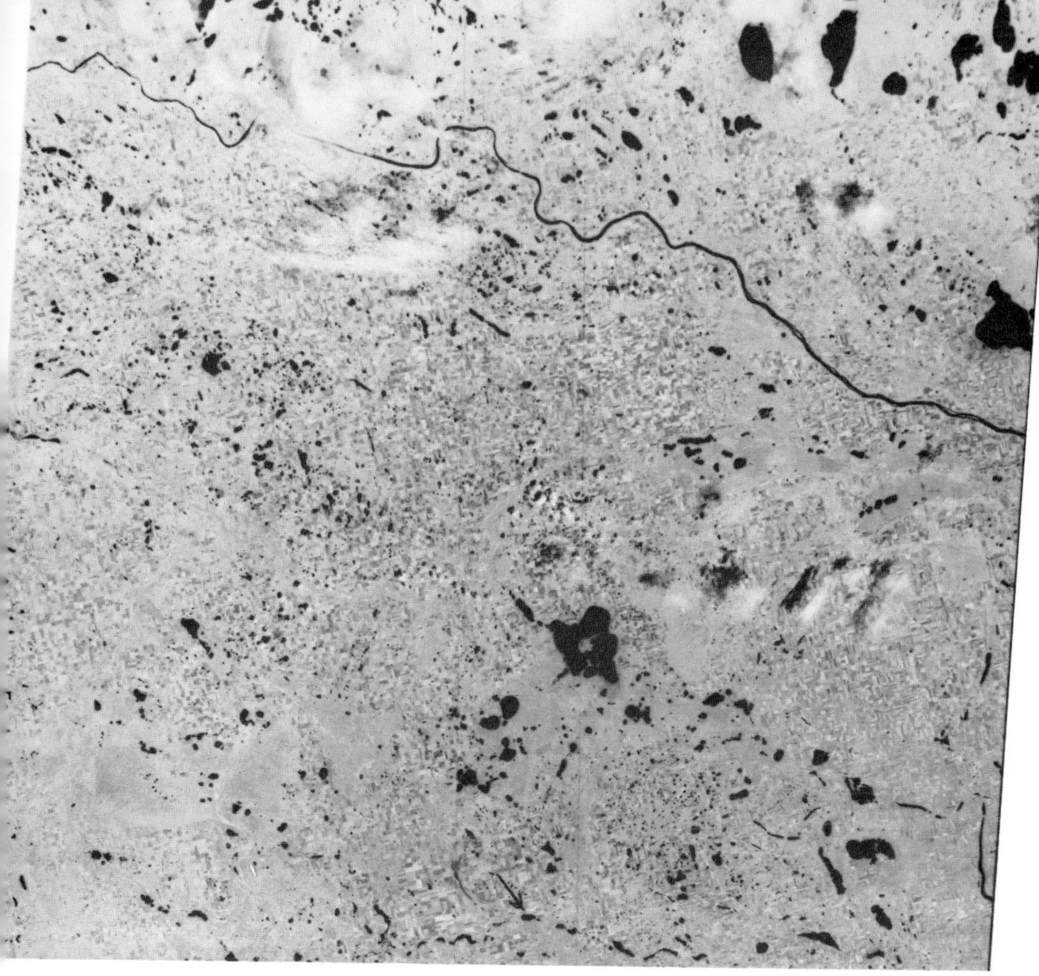

Fig. 7.14. Image of west-central Saskatchewan, Canada, 9 August 1972, from Multispectral scanner Band 7. (NASA ERTS E-1017-17494-7 01.)

characterized by large springs where the present valley bottom is situated below the base of the aquifer. Water wells in such areas often yield 2000–5000 litres per minute. Other aquifer sites that can be identified on satellite images include chains of small elongated lakes located at intervals along abandoned meltwater channels, and large alluvial fans where major tributaries discharge into lakes or into former meltwater channels. An example of the chain of elongated lakes can be seen at the bottom of Fig. 7.14 running almost east and west. One of these, at the point marked with an arrow, has been found to be deep, free of detrimental salts and algae, and bordered on the south by deep outwash sand and gravel, which is water-bearing. Two wells in this area supply water to the towns of Provost in Alberta and Mackin in Saskatchewan, Canada.

One other interesting feature in Fig. 7.14 is a small lake which shows up conspicuously on both bands and shows a white crescent-shaped 'halo' on its southern shore in Band 5. It is indicated by an arrow in Fig. 7.13. Springs in the depression in which this lake is located are tapped to provide nearby municipal water requirements. The 'halo' is produced by long-continued evaporation of groundwater discharge and consequent precipitation of alkali salts.

Snow and ice

More than three-quarters of all the fresh water in or on the earth or in the atmosphere above the earth at any given time is 'in storage' in the form of snow and ice. This is a tremendous reserve, although much of it is in glaciers and ice sheets, only a small part of which is formed and released seasonally. The seasonal snow cover, which blankets about a quarter of the land surface for more than four months each year, is of critical importance to stream flow in many areas. In some, virtually all of the stream flow is derived from melting snow.

Snow cover

Snow can be readily distinguished from bare land in *Landsat* images (see Fig. 7.9) and also in good-quality meteorological-satellite images. It can usually be distinguished from clouds by lack of motion or change in pattern. Forested terrain presents some difficulties, but if a forested area is surrounded by or contains snow-covered clear areas at the same elevation, we can reasonably surmise that there is snow under the trees. There are some situations where bare rocky surfaces fully illuminated by the sun will appear brighter than snow cover in a shadowed valley, but a reasonably experienced viewer can usually make the distinction with little difficulty. For routine measurements of snow cover to be of operational value, however, the process must be automated. Hand measurement with a planimeter or by dot-counting is too time-consuming. Also, the hydrologist needs to know not the total snow-covered area in a complete *Landsat* image but rather the area in various specific drainage basins. Procedures using an interactive electronic-computer console, have been developed, in which a drainage-basin outline map is superimposed and registered on a portion of a *Landsat* image. The operator, using a cursor, delineates the particular area in which he desires to know the snow cover. He then selects a brightness level using one or several spectral channels that seem best to distinguish snow from other background,and rejects all areas below that level. He brightens shadowed or vegetation-covered areas and darkens areas that appear, by comparison with images taken at other times, to be clouds. After the snow-covered area has been thus delineated and defined, it is a simple matter for the computer to measure its total area in the particular drainage basin concerned. Accuracies of about 96 per cent can be achieved.

By comparing the *Landsat* image with topographic maps, the altitude of the local snow line can be found with an accuracy of about 60 metres; or alternatively an 'equivalent' snow line, defined as the altitude above which the area in the drainage basin is equal to the measured area of snow cover, can be determined. Snow-line altitude is related to the level at which the temperature does not rise above freezing for any significant period each day.

Experience shows that seasonal run-off volumes are highly correlated with snow depletion rates and that the general form of this relationship remains much the same from year to year even though the total amount of snow and the weather conditions may vary. Therefore periodic measurements of snow-covered areas can be quickly converted to snow-pack storage estimates and stream-flow forecasts. These forecasts can be further improved by combining the snow-cover data with meteorological information and a limited number of measurements at remote ground stations relayed through the *Landsat* data-collection system. The monetary value of such forecasts is significant. A single electric utility system in the north-west United States reportedly realized savings of about a million dollars (US) during the first year that data from four new snow courses were incorporated in its operational forecasts. There are also potential ecological benefits, because accurate forecasts of stream flow and reservoir inflow make it possible to avoid excessive or deficient outflow.

Glaciers

Glaciers, which are slowly moving 'rivers' of ice, accumulate from compressed snowfall at high altitudes and gradually move through mountain valleys to lower altitudes where they melt, or in colder climates come to the sea coast where they break off to form icebergs. Snow lines on glaciers are not difficult to identify in *Landsat* images. Thus it is possible to determine the accumulation area ratio for a large number of glaciers rapidly and economically. This is important because the accumulation area ratio is a useful index to the mass balance of a glacier. By measuring the distribution of ratios in a region and relating these to meteorological measurements at a single glacier research station in the same region, it is possible to extend the meteorological data over the entire region, a necessary step in understanding glacier meteorology, which has hitherto been difficult to achieve.

The *Landsat* images show in remarkable detail what are probably faint dust bands and medial morains on the ice that disclose the directions of ice flow. These markings are of such a large scale as to be essentially unrecognizable on the ground or from aircraft. One example is an interesting pattern recognized for the first time in *Landsat* images of the Bering Glacier in Alaska. Just below the summer snow line, the ice from the two main snow fields which feed this glacier merges from east and west and then bifurcates—the Bering Glacier flowing south-west and the Tana Glacier splitting off from it to the north-west. The situation is further complicated by the fact that the Bering Glacier surges—that is, it moves in slow pulses—whereas the Tana Glacier seems to flow at a more or less steady rate. Markings that can be seen in the *Landsat* images indicate that during the surge phase all of the ice from the western field is channelled into the Bering Glacier. When this lobe of ice begins to move into the Bering Glacier, a rapid advance of the terminus 80 kilometres away can be expected to occur about three years later.

Surging glaciers can generally be identified and monitored by means of satellite imagery. What causes a glacier to surge and why some do while others do not is of considerable scientific and practical interest, because this type of periodic sudden slippage is common to other natural phenomena such as earthquakes. Surging glaciers may advance over large areas and cause devastating floods by blocking and suddenly releasing large quantities of melt water. Forecasting the forthcoming outburst of the large Berg Lake and other likely changes in drainage will depend on the changing flow characteristics of the Bering Glacier.

Sea ice

Sea ice is primarily of interest in the Arctic Ocean, Baffin Bay, and around the shores of Antarctica. It plays an important role in control of climate not only in these regions but all over the world. It also acts as a barrier to shipping in and out of ports on the northern shores of Alaska, Canada, Norway, Finland, and the Soviet Union. Some information about the dynamics and morphology of sea ice has been obtained over the years by adventurous scientists who set up research stations on floating ice islands in the Arctic, where they lived and worked the year round. More recently, such measurements have been supplemented by aircraft equipped for remote sensing with photographic, infrared, and microwave devices, by data from the *Nimbus* meteorological satellite, including the tracking of radio transponders on drifting buoys that are sometimes frozen into the ice, and by *Landsat* images.

Only the *Landsat* images have sufficient resolution to show individual floes, leads, and polynyas and to enable estimation of the magnitude and direction of strains in the ice pack. But unfortunately the polar regions are dark half the year, and during this period *Landsat 1* is of no value. The thermal infrared channels of *Nimbus* imaging devices (described in Chapter 5) can operate in the dark, and the heat flux from newly formed open leads in the ice is as much as 100 times greater than that of the surrounding ice. But the approximately one-kilometre resolution of these devices does not show small features as well as desired. The scanning microwave radiometer ESMR on *Nimbus* (also described in Chapter 5) has a resolution of about 20 kilometres and therefore cannot show small features at all. It can, however, distinguish between open water, new ice, and multi-year ice by the difference in emissivity, which is affected by brine inclusions and crystal structure, and it can also cover the entire area of the Arctic Ocean (or Antarctica) every day. Consequently the ESMR has considerable value for large-scale studies. For smaller-scale studies, a combination of photographic, infrared, and microwave imaging equipment on high-flying aircraft still seems to be required. Such a combination has been used, together with *Nimbus* tracking of drifting buoys and *Landsat* images when available, in the Arctic Ice Dynamics Joint Experimental Programme (AIDJEX), which took place during 1970–3 in selected areas of the Beaufort Sea.

Satellite measurements of sea ice will be improved in the future by the addition of a thermal infrared sensor channel on another *Landsat,* and perhaps ultimately by the development of microwave imaging devices of higher resolution.

The oceans and the seas

The 70 per cent of the earth's surface that is covered by salt water—the oceans and seas—represents a major challenge to scientists and explorers as well as a vital resource of food and minerals for mankind. Until about 25 years ago it was studied only by means of a few individual research vessels, rather crudely equipped in terms of today's technology. During the 1950s and 1960s, oceanographic instruments and techniques were greatly improved, and coordinated projects involving several ships and instrumented aircraft were carried out with interesting and useful results. Nevertheless, the vastness and dynamic character of the oceans has inevitably caused oceanographers to turn toward satellite technology as a new and powerful means for extending their studies.

Light in the yellowish–green wavelength region of *Landsat* MSS Band 4 decays to about a quarter of its surface value at a depth of 10 metres in relatively clear sea water, whereas the near infrared radiation to which Band 7 is sensitive can penetrate only to a depth of a few tens of centimetres. Bands 5 and 6 penetrate to depths intermediate between these limits. Band 5, for example, penetrates only to about a quarter of the depth of Band 4; however, Band 4 suffers more from atmospheric scattering, so that for the same surface signal, contrast is reduced by a factor of almost 2 when viewing in Band 4 as compared with Band 5. For some purposes, therefore, Band 5 is preferable.

Shoreline information needed to update nautical charts and maps is most clearly delineated in Band 7, whereas shallow underwater features such as shoals and reefs

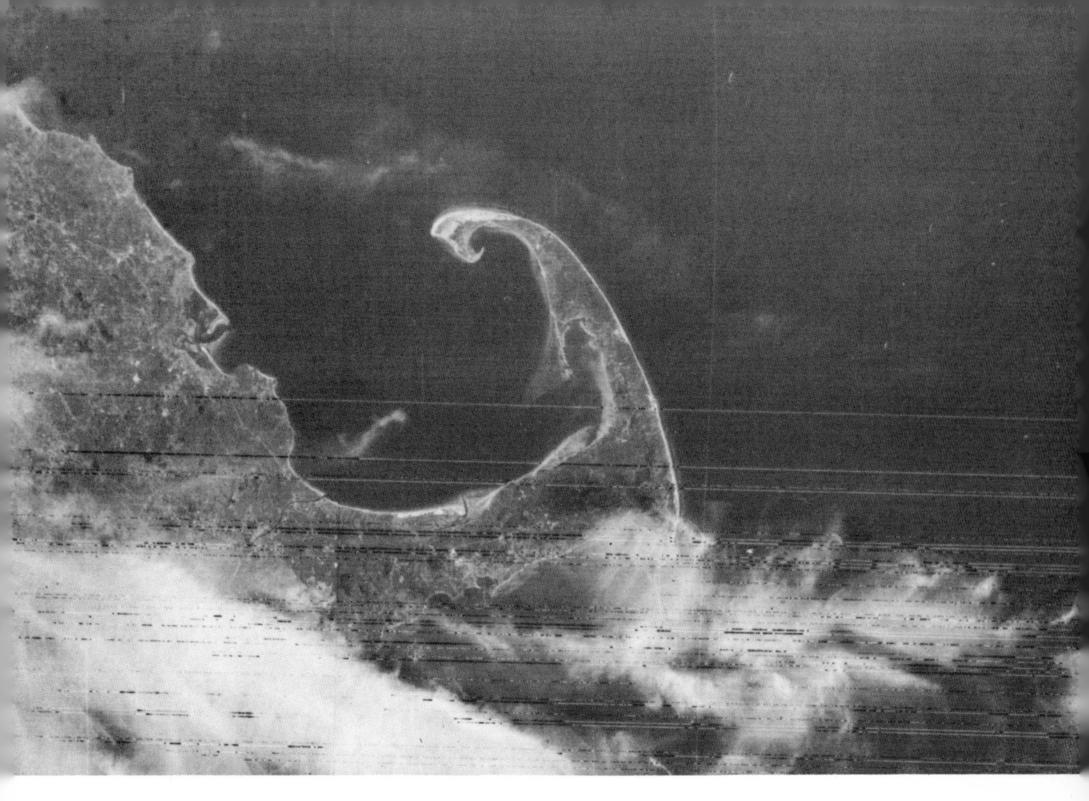

Fig. 7.15. Image of Cape Cod, USA from multispectral scanner Band 4 (NASA ERTS E-1040-14522-4 1.)

Fig. 7.16. Image of Cape Cod, USA from multispectral scanner Band 5 (NASA ERTS E-1040-14552-5 1.)

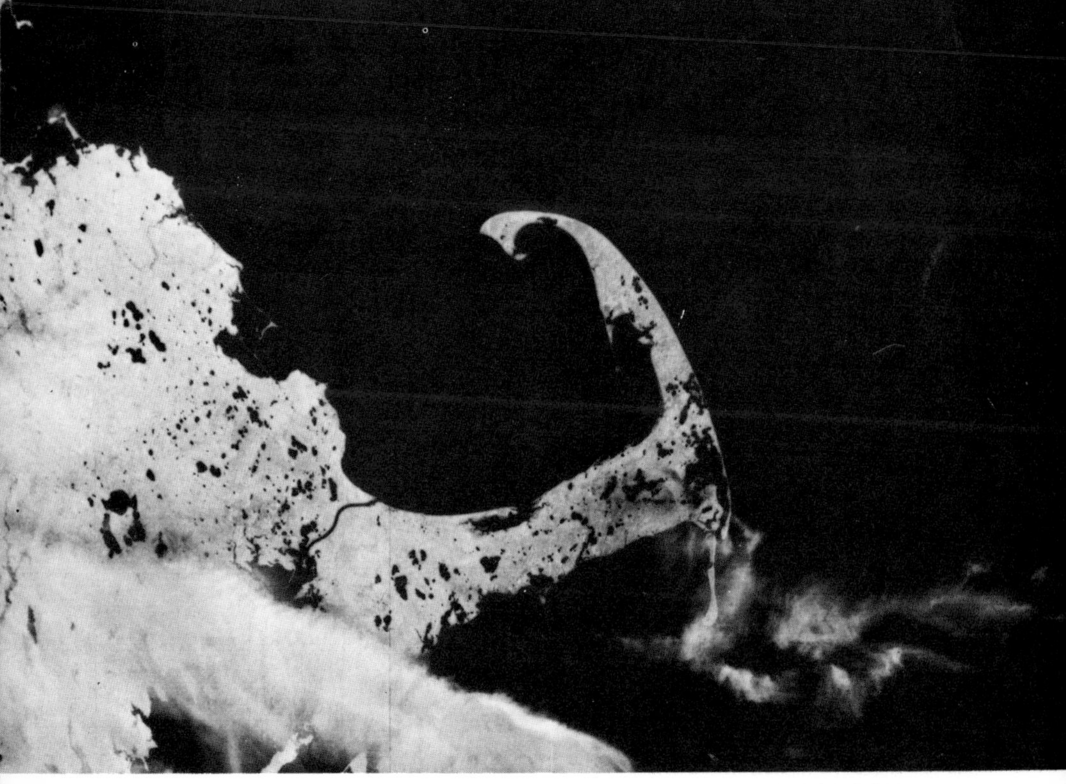

Fig. 7.17. Image of Cape Cod, USA from multispectral scanner Band 7 (NASA ERTS E-1040-14552-7 01.)

can be obtained from Bands 4 and 5. Figs. 7.15, 7.16, and 7.17 show images in these bands for the Cape Cod (USA) area made in September 1972. Note, for example, Barnstable Harbour, which lies almost directly south of Provincetown on the northern tip of the Cape. In Band 7 (Fig. 7.17) this looks like a large well-protected harbour with room for many ships. In Band 5 (Fig. 7.16) it can be seen that most of it is shoal, though there does appear to be a dredged channel with a moderate-sized basin at its end. In Band 4 (Fig. 7.15) the shoals extend farther out into Cape Cod Bay and the basin shrinks to almost nothing, although the dredged channel can still be seen. The Cape Cod Canal, which is very well defined in Band 7, is barely discernible in Band 5 and probably will not be seen at all in this reproduction of the Band 4 image. Note also how the obscuration of the thin clouds decreases from Band 4 to Band 7. From similar images over the Bahamas banks correlations have been made with measured depths at 2, 3, and 8 metres, and a mathematical model relating depth measurements to *Landsat* MSS Band 4 and 5 image density ratios has been developed. Where conditions of reflective bottom and clear water occur, density contouring of Band 4 images alone can be used to estimate depths in 2–, 5–, and 10–metre steps. Thus it is already apparent that *Landsat* images can be used to provide up-to-date information about navigation hazards, especially shifting sand bars, in remote areas of the world.

By the use of contrast enhancement and density contouring, it is also possible to detect areas of increased turbidity in oceans, seas, and estuaries. These can generally be distinguished from shallow reflective bottom areas by colour, as indicated by the ratio of Bands 4 and 5. To a certain extent, the nature of the turbidity, whether caused by fine silt or biological organisms (chlorophyll, for example) can also be determined by comparing different colour bands. Turbidity caused by silt is a good

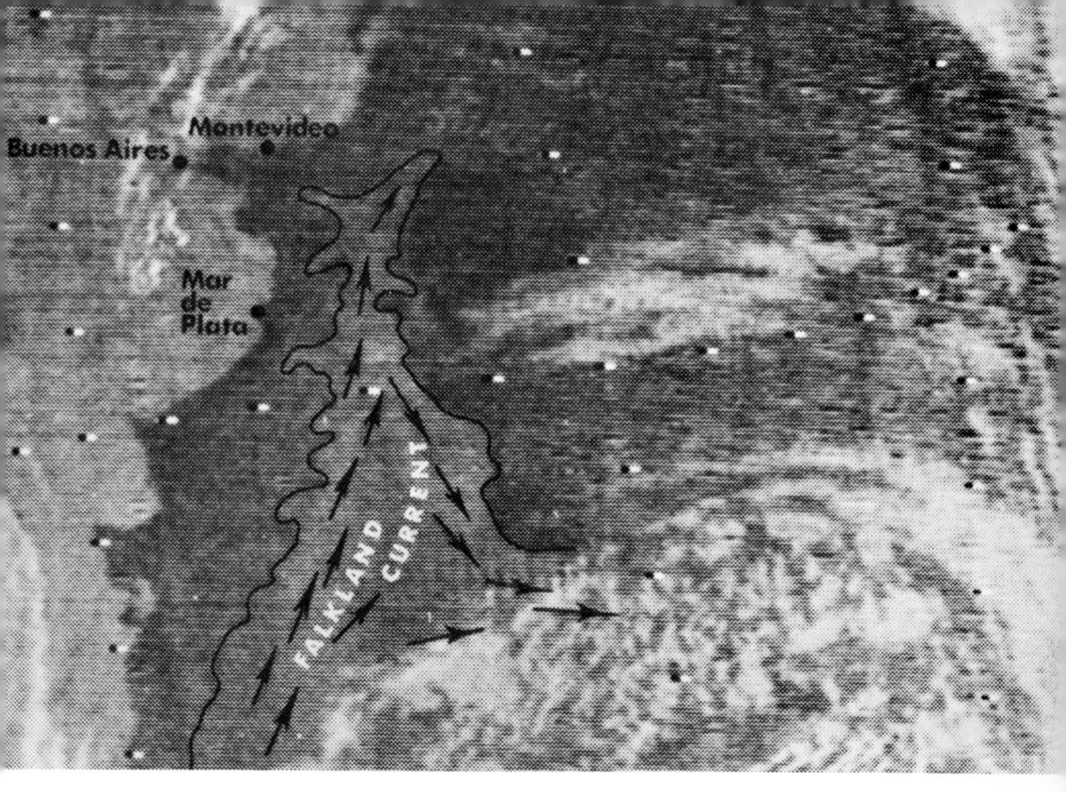

Fig. 7.18. *Nimbus* HRIR image of east-coast South America showing Falkland Current.

natural tracer for use in observing water movements in estuaries and bays. It is especially useful for identifying the boundary between fluvial (river) and marine waters. Such boundaries are important because pollutants are often associated with turbid water, because fish tend to congregate at such boundaries, and because logs and other floating debris which constitute hazards to navigation tend also to accumulate there. In open ocean areas increased turbidity is likely to be associated with biological activity resulting from nutrient-rich water upwelling from the ocean floor and is generally associated with areas of low surface temperature that can also be observed by thermal infrared measurements from meteorological satellites, and later will be observable with better definition by the thermal infrared channel to be incorporated in *Landsat*. Such measurements have already proved useful in locating areas of potentially high yield for commercial fisheries in the Gulf of Mexico and elsewhere.

Large-scale currents

Large-scale ocean currents can be observed and monitored by measurement of the surface temperature differences across the stream boundaries using thermal infrared images from meteorological satellites. Much of this work has been concerned with the warm Gulf Stream, which flows north-east along the eastern coast of the United States and thence across the north Atlantic to Europe, where it helps to maintain a milder climate than would otherwise be expected at these high northern latitudes. A recent study at the Instituto de Pesquisas Espaciais in Brazil used 30 relatively cloud-free images from a *Nimbus* 11·5-micrometre infrared channel to chart the seasonal movements of the boundary between the cold Falkland Current and the warmer Brazil Current. This boundary was observed to move as much as 480 kilometres over the course of a year. Fig. 7.18 shows an infrared image of this current.

The warmer coastal water near the mouth of the Rio de la Plata, the colder Falkland Current off shore, and the sharp boundary between this cold water and the warm Brazil Current still further off shore can be seen clearly.

Studies of the Loop Current in the Gulf of Mexico indicate that the edge of this current is an accumulation zone for surface algae and phytoplankton which exhibit high reflectance in *Landsat* Band 6. In addition there is frequently a difference in sea state across the current boundary, steeper waves and more whitecaps occurring in the current, which therefore appears brighter in *Landsat* Band 5. In Band 4, however, the bluer water in the current appears darker than the surrounding greener water of Florida Bay.

In a much more quantitative sense, ocean currents can be measured by using satellites to track free-floating buoys or to collect data from moored arrays of instruments. The former are sometimes referred to as Lagrangian current measurements, after the dynamical analysis method of Lagrange, in which individual particles are identified and their behaviour followed as time runs on. These can be made using either surface buoys or submerged buoys designed to remain at a constant pressure depth, surfacing only briefly at intervals of between two and four weeks to make their location known and then submerging again. The second type of measurements are called Eulerian, from Euler's approach to dynamics which describes fluid motion in terms of its behaviour at various fixed points in space as a function of time. Magnitude and direction of current flow are measured at a number of fixed locations on the surface of the ocean and also at various depths, and the data are collected by satellite or other means of communication. Both techniques have already been established and will continue to be used for detailed dynamic study of limited sea areas by individual research groups. Tracking of a group of 10 or 20 buoys simultaneously released at a single point, or of several such groups at different points, is probably the most economic and productive approach. Tracking of buoys by ships or aircraft is very costly in terms of ship days or flying hours, and consequently satellites can make an important contribution in this area.

Wave studies

The study of waves, ranging from capillary and small gravity waves to larger locally generated waves and ocean swells propagated from distant regions, is a complex but important subject. Laser and narrow-beam radar altimeters carried by low-flying aircraft have been used to measure detailed wave motion, but it seems unlikely that this sort of measurement can be extended to satellites. A more promising use of satellites would seem to be for measuring one or two parameters representing 'sea state' over an area of a few hundred square kilometres and a few minutes to a few hours time, and especially for determining the global distribution of these parameters on a daily basis. A statistical measurement of surface roughness can be used as a parameter, for example, and this can reasonably be taken as an indicator of wind speed with application to the generation of ocean-current circulation and as an input for meteorological modelling.

Most of the requirements for wind-wave sensing are embodied in an instrument package designated as S-193, recently tested in the United States Skylab satellite. This combination, which is shown in Fig. 7.19, can function as a scanning narrow-beam radar, a passive microwave radiometer, and a radar altimeter. The back-scattered radar return, as a function of angle from the vertical, can be interpreted in

Fig. 7.19. Photograph of S-193 radiometer–scatterometer instrument package used on *Skylab*.

terms of roughness. The passive radiometer signal depends partly on roughness and partly on sea surface temperature. Both types of measurement are affected to some extent, although in different ways, by heavy cloud cover and precipitation. Under at least partly clear conditions, temperature corrections can be derived from infrared sensors also included in the spacecraft. If the temperature is known at least approximately, anomalies due to clouds, precipitation, or foam, sensed by the radiometer, can be used to correct the radar data or in extreme conditions as a warning to disregard it. The wavelength used is approximately 3 centimetres. Thus the data apply primarily to very short-wavelength components of the sea wave spectrum. Amplitudes of the longer waves, which in fact transmit most of the energy, can be inferred from the theoretical shape of the wave spectrum, at least for steady wind conditions; however, this theory is somewhat speculative and ignores swell produced by storms and wind systems at a distance from the area being observed. Consequently an independent means is desired to sense a parameter that adequately represents wave height. Such a parameter can be obtained from the shape or 'spread' of the return pulse from the radar altimeter. An instrument package like the S-193 could potentially generate data equivalent to 40 000 ship reports per day over all the ocean areas of the world.

The surface of the ocean

The detailed dynamic shape of the ocean's surface, expressed in terms of distance from the centre of the earth as a function of latitude, longitude, and time, is of great interest to physical oceanographers. It is, however, very difficult to measure by ordinary methods; even space technology is not yet well enough developed to provide data of the desired accuracy, although there is good reason to believe that it will be able to do so in the future. The largest variations, of some tens of metres, are caused by gravitational anomalies or distortions in the shape of the geoid. These were mentioned on p. 100. The surface of the ocean departs from the geoid by distances of the order of a metre because of the 'tilt' that is induced by large current systems and to a lesser extent by variations in density of the sea water related to variations in temperature and salinity. Such departures are almost constant with time but can be distinguished from geoidal variations by using sensitive shipboard gravitational measurements to determine the geoid in interesting locations. Measurements of this kind represent one of the few ways by which the integrated internal motions of the oceans can be measured.

Variations of the surface height with time are, of course, independent of the geoid and can rather easily be distinguished from the latter if sufficient data are available. The most important of the variations with time are the tides caused by the gravitational attraction of the moon and sun. Tidal amplitudes are of the order of a metre over most of the ocean area and the variations are strongly periodic, so they can easily be measured even in the presence of other variations or 'noise' of about the same amplitude. Remotely sensed tides are truly geocentric, that is, they are measured with respect to the centre of the earth rather than with respect to the earth's crust. This is an advantage because the crust itself is deformed by about 0·2 metres by the gravitational forces exerted by the sun and moon, and also by about 0·1 metres, in some places, by the variable stresses produced by the weight of the ocean water as the tides rise and fall. Other variations of ocean surface height with time that might usefully be measured would be those caused by changes in barometric pressure and by winds.

In order to make useful measurements of these variables, it will be necessary to achieve an overall accuracy somewhere between 0·1 metres and 1 metre for the satellite-tracking and radio-altimeter system. An appropriate bias for the effect of waves on the altimeter measurements under different sea conditions must be derived and used. Attaining this order of accuracy is certainly a formidable goal; however, with continued improvement of the techniques that are now being tested in the geodetic satellite *GEOS C*, it seems likely that this goal can be reached.

8 Monitoring flora, fauna, and fish

Be fruitful and multiply, and fill the earth and subdue it, and have dominion over the fish of the sea and over the birds of the air and over every living thing that moves upon the earth
The Bible, Genesis I-28

A glance at population statistics will show that mankind has already rather well 'filled the earth' and is now predictably on the way to overfilling it—probably with disastrous consequences! And it would appear that man has exerted his 'dominion over the other living creatures in such a careless and ignorant way that some of these creatures have already disappeared from the face of the earth and others seem very likely to follow. Clearly man is no longer in the Garden of Eden. Unless he can learn to limit his own numbers, to produce food more efficiently and abundantly, to use natural resources more wisely, and to live in greater harmony with his environment and with the other living things that share it, his future will be bleak. These problems, perhaps the most important of our time, are the proper concern of biology, fishing, agriculture, forestry, demography, and anthropology. In this chapter, an attempt will be made to examine some of the ways in which satellite technology can contribute to progress in such areas.

Satellites and the biologist

In the past biologists, with rare exceptions, have been accustomed to looking at the individual animal rather than the herd, the tree rather than the forest. Therefore until recently satellite technology had little application in biology, except as a means to observe the effect on living organisms of the almost zero gravitational field that prevails in a satellite environment, and as a way to eliminate all phenomena associated with the earth's daily rotation in studies of the mysterious biological rhythms exhibited by most terrestrial creatures. Recently, however, as a result of increased interest in ecology and perhaps more specifically because of the International Biological Programme, biologists have been learning to study large complex natural units called 'ecosystems'. An ecosystem is described as a community of interacting species that can be differentiated from other communities. This description includes communities as small as a colony of micro-organisms on a single leaf, but it usually refers to larger, more permanent systems such as a lake, a forest, an island, or a region of desert or Arctic tundra. Satellite images and spectral measurements can be of great value in this kind of biological work and are being increasingly used for it.

Signals of biological interest are found in many parts of the electromagnetic spectrum. The wavelengths of visible light used in ordinary colour photography are of obvious significance. The near infrared, covered by *Landsat* Band 7, is of great importance in plant studies, because healthy green plants have a high reflectance in this region whereas bare soil and dead plants have a low reflectance. Plants or crops injured by disease, insects, drought, or improper soil conditions show intermediate reflectance values. The medium and far infrared wavelengths are also useful in studying soil and crop conditions and in the study of animal life on land or in the sea. Healthy vigorous vegetation with high transpiration tends to be cooler, by as much as 10 °C, than the surrounding air or soil in the daytime and slightly warmer at night. Diseased vegetation can show abnormal temperature response. Herds of large terrestrial animals are warmer than their surroundings at night. Fish shoals can sometimes create discernible variations in water temperature because their movement mixes water from strata having different temperatures. A very obvious application for thermal infrared satellite sensors is in the location and monitoring of forest fires, a job now being done routinely by aircraft in many areas.

Some biologists have long wished for effective means to follow the movement of individual birds, land animals, and sea animals, especially those that migrate seasonally or at critical points in their life cycle. This requirement could theoretically be met by attaching small radio transponders to the animals, which could be located and identified periodically by satellites. At present, however, it does not seem to be technologically feasible to design a transponder and power supply with sufficient power and operating time in a package small enough and light enough to be carried by any but the largest animals. Some experiments have been carried out but so far without great success. Future advances in technology may improve this situation.

Biological studies in the Antarctic

The oceans surrounding the Antarctic continent are largely covered by annual sea ice which extends north in winter and retreats to the south by melting in summer. This region is referred to as pack ice and is biologically rich because ocean currents converge here in a manner that creates upwelling and high nutrient content, which supports biological activity at all levels from plankton to whales. It is an inhospitable region for man, and consequently the magnitude and extent of the biological resources are poorly known. There is, however, considerable interest in Antarctic seals, whales, and other vertebrates, as well as in some invertebrates; therefore biologists are looking more and more to remote sensing from aircraft and satellites to aid them in the assessment of these resources and in better understanding the Antarctic pack-ice ecosystem. Naturally life is more visible during the southern summer when the region is exposed to continuous sunlight. During this period phytoplankton multiply rapidly and the vertebrate populations go through their annual reproduction cycle. Thus the visible vertebrate densities are greatest in late March, because the pack-ice area is smallest and the highest population levels in the yearly cycle then exist.

Penguins and seals

Each species has different characteristics with respect to its susceptibility to remote sensing. Penguins, for example, congregate on rookeries to nest and breed. During this period colony sizes vary from a few thousand to more than 100 000 birds. The rookeries, which are generally located on outcrops of land and occupied year after year, are large enough and have enough contrast to be observed in *Landsat* images. Any change in activity from year to year can be estimated from the size and brightness of the annual deposit of guano. Crabeater, Ross, and leopard seals have their pups on the floating pack ice and do not congregate to any significant extent. However, Wedell seals tend to occupy heavy pack-ice regions close to the continent and tend to congregate in large numbers along annual tide cracks to breed and have their pups. Because of their high contrast, breeding colonies of Wedell seals may possibly be detectable in images of *Landsat* quality.

It is obvious that the resolution of present *Landsat* images is too coarse (by about two orders of magnitude) to show individual seals. The necessary resolution can be obtained using aircraft but surveying the pack-ice region is a difficult and expensive task with the types of aircraft currently available. It could also be obtained by developing satellite sensors with increased resolution. There would be no inherent difficulty in achieving a resolution on the order of one or two metres, as far as the optical design is concerned, if the required space and weight could be provided in the satellite. The principal limitation has to do with data handling. An increase of two orders of magnitude in resolution would generate 10 000 times more data per unit area to be surveyed and would create a truly formidable problem unless the surveyed area were to be decreased in proportion. There would also be more demanding requirements on attitude stabilization and motion compensation, although these could probably be met without great difficulty. In any case, it does not appear that any significant fraction of the pack ice can be surveyed with an image resolution sufficient to show individual animals unless a much larger and more expensive satellite than *Landsat* were to be developed especially for the task.

In view of the difficulty of high-resolution imaging, attention is being given to other methods, such as the use of hydrophones suspended from air-dropped buoys or implanted by icebreaker ships during the summer. These could be used to detect the movement of individual seals tagged with sonic devices or even of untagged animals by listening to their vocalizations. To be effective, many such sensors would be required; however, they could be designed to operate for many months, at least through the winter season, and need not be very expensive. Recordings would be routinely read out by satellite.

Krill

Probably the most interesting and important living creature in the Antarctic marine ecosystem is the krill (*Euphasia superba*), a small planktonic crustacean about 5 centimetres in length and having an average weight of about 1·2 grams at maturity. Krill are the principal food for many species of whales and other marine animals. Standing stocks of krill have been variously estimated at about 150 million to 950 million tonnes, well in excess of the total harvest of all fish from all the oceans of the world. With the demise of the stocks of baleen whales in the past 40 years and serious decreases in other whale species, attention is now being given to the possibility of harvesting krill themselves as a source of high-protein food for the hungry world. Interest in the study of krill is therefore not wholly academic; man can ill afford to decimate the krill population as he has done with other valuable marine resources.

The growth and distribution of krill are known to depend on the abundance of chlorophyll-containing phytoplankton, water temperature, current, amount of sunlight, and proximity to the edge of the pack ice, all of which can be monitored by satellite instruments. Methods have not yet been developed for direct observation of krill from satellites; however, it does seem likely that such observations will be found to be feasible with no great stretch of technology. Krill vary in colour from brilliant red to rust or pale yellow. Swarms, which average 40 metres by 60 metres in size (maximum 600 metres), are generally found at depths of 10–40 metres during the day and 0·5 metres at night. They are dense enough to discolour the sea water and can easily be spotted from the deck of a ship. Patches of this size are frequently spaced about half a kilometre apart over an area covering 150 square kilometres. In addition to having a distinctive colour and swarm morphology, krill are bioluminescent. Swarms of *Euphasia pacifica,* a related crustacean, have been photographed from aircraft at night off the coast of California using low-light techniques. There seems to be no reason why such techniques could not also be used from satellites. Unfortunately, cloud cover over the southern ocean areas inhabited by krill is of the order of 50–90 per cent. However, breaks do commonly occur, so with patience and a bit of luck, reasonable sample observations may be obtained.

Fish and fisheries

As the world population has increased, the harvest of living marine resources has likewise increased; in fact it has grown at a more rapid rate than the human population. During the last few years, however, it appears to have tapered off somewhat, giving rise to a concern that the already significant animal protein 'gap' in world food supplies may widen. It is therefore vital to determine, as accurately as possible, the ultimate potential that can be expected from the world's fisheries and how best to achieve this potential. On the one hand, it is important to understand the influence of both the environment and of human activity on the population size of useful species so that effective means, including limitation of harvest, can be established to keep these population sizes at optimum levels. On the other hand, it is desirable to devise more efficient means to find and harvest fish so that, within any restrictions that may have been established, the cost of the catch in terms of man-hours, fuel, and other resources may be minimized.

It would be particularly useful if, through the use of satellite technology, the standing stocks of all useful fish could be accurately evaluated at intervals of, say, a year. Unfortunately, such an objective does not now seem to be within reach. There is very little possibility that satellite imaging systems will ever be capable of surveying large areas with fine enough resolution to 'see' individual fish or animals, or that, even if this were possible, anyone would know what to do with the resulting data. And, of course, there is no possibility at all of seeing fish at a depth greater than a few metres.

Locating schools of fish

Schools of fish feeding at the surface are slightly more promising targets, ranging in size from about 10 metres to about 1000 metres across. However, most sea creatures have evolved protective coloration and hence have low contrast with respect to their surroundings at visible or photographic wavelengths. Fish have also developed low body temperatures or efficient thermal insulation in order to avoid losing excessive amounts of energy to the water in which they live, and so they show little contrast in the thermal infrared part of the spectrum. It seems likely that schools of fish will be most easily detected by the characteristically active movement they exhibit when feeding. In a calm sea, the surface would be sufficiently disturbed to change its reflectivity significantly, and this could probably be seen in *Landsat*-type imagery if the school were large enough. Mixing of the thin layer of surface water with water of a different temperature just below also produces a thermal wake behind a school of fish, which can be seen by a sensitive infrared scanner. Measurements of microwave radio emissivity as a function of wavelength, using a multi-frequency radiometer or scatterometer, might even provide a species-specific signature of the surface roughness caused by a school of fish if the geometric resolution could be made fine enough. Similarly, it has been suggested that the oil slicks produced by large schools of fish when they feed could be observed from a satellite by their effect on surface roughness, and could be differentiated from other kinds of oil slicks by ultraviolet spectroscopy. Unfortunately, this scheme has not been successful because of problems with atmospheric scattering at ultraviolet wavelengths.

Many small ubiquitous marine organisms, particularly dinoflagellates, are bioluminescent when stressed by the turbulent wake of ships or by the thrashing about of feeding fish. This bioluminescence, which takes the shape of the school of fish, can be seen visually from the deck of a ship or from aircraft. Pilots who are experienced fish-spotters can identify more than a dozen different species of fish by the different patterns of light flashes emitted from schools of these species. Low-light-level television cameras with image intensifiers have produced and recorded good night-time images of fish schools from aircraft at about 2000 metres. Presumably this could also be done from satellites.

Even if schools of fish could be reliably detected and identified in satellite imagery, however, the task of scanning the data appears to be extremely difficult. Yellow fin tuna, for example, are distributed over a 22 000 000 square kilometre area in the eastern Pacific Ocean. This would require about 640 *Landsat* images (185 kilometres by 185 kilometres) to cover. There would probably be about 10 000 schools of fish in this area. An individual school of yellow fin tuna having an average size of about 250 metres would be represented by a speck about a quarter of a millimetre in diameter and, allowing for the fact that only about a third of the schools would be feeding on the surface at a given time, there would be only about five of them on the average in each *Landsat* image. It therefore seems likely that sampling procedures would have to be used, although they will be of questionable validity because of the typically non-uniform distribution of the fish.

Predicting where the fish are

What satellites can do is to locate with some degree of dependability the areas in which fish are likely to congregate. Because they are continuously moving about in search for food, tuna fish are likely to respond to environmental factors that result in greater food abundance, such as the presence of oceanic fronts, where forage tends to be concentrated. Fishermen have long been aware of the ability of fish to modify their distribution in accordance with temperature changes in the surface water. Sea surface-temperature maps showing the location and character of fronts are now being prepared from satellite infrared sensor data as a possible means of improving sampling procedures and as a practical aid to commercial fishermen. More recently the depth of the thermocline (region of steep vertical temperature gradient) has also been recognized as having an influence on the distribution of fish. Tunas seem to prefer swimming above the thermocline, possibly because density gradients associated with the thermocline may cause plankton and other organic matter to collect there.

Although it is not possible to measure thermocline depth directly by means of satellites, models exist from which this parameter can be estimated as a function of wind intensity and duration, air temperature, and sea surface temperature, all of which can be obtained with sufficient accuracy from satellite meteorological data. Sediment outfall is another indicator that is useful in some kinds of fishing; as previously noted, it can be observed in satellite images. Curiously enough, persistent cloud cover, usually regarded as an obstacle to satellite observations of the ocean surface, may indicate the location of a current and associated areas of good fishing, as has been demonstrated by the French Research Vessel *Calypso* using APT images of clouds over the Peru Current.

Satellites and fishermen

Weather data from satellites has great practical utility to fishing fleets and is already widely used. Satellite navigation and communication will probably be more widely used in the future. For enforcing international fishing treaties, it may be necessary to require that all seagoing fishing vessels be registered internationally and carry a small inexpensive radio transponder similar in principle to those used for tracking balloons from satellites. In this way, it would be possible to record the time spent by each ship in any given fishing area and, with some knowledge of the ship's capability and habits, estimate the magnitude of its catch.

Agriculture and forestry

Recent shortages of food, lumber, and paper underline the importance of better planning and management of farming and tree-growing activities. Plants and trees grow more vigorously and produce better yields in the right soil with the right amount of moisture and when not adversely affected by diseases and pests. As more agricultural land is taken out of production each year by 'urban sprawl', as more marginal land is put into production, and as more highly specialized plants and animals, bred for high production but possibly more susceptible to certain pests and diseases, are grown, the natural ecology becomes increasingly disturbed. Very wise management will be needed to avoid catastrophe.

Soil maps

Detailed soil maps at large scale are made by walking over the land and boring holes. Smaller-scale maps, usually at 1:500 000 or 1:1 000 000, called soil association maps, are used for broad planning purposes in civil engineering as well as in agriculture. They are not as precise as large-scale maps but they cost much less to make, cover a much larger area, and are field-checked only at infrequent intervals. The different soil 'landscapes' that are identified on these soil association maps can be characterized by surface geometry expressed in streams and relief features, type of vegetation for various climates, hydrology, and soil composition or colour. All of these characteristics can be observed in *Landsat* images, and the scale of the images normally produced is about 1:1 000 000—just right for this purpose. Soil maps have been made from *Landsat* images not only for a number of areas in the United States, where previous information was available, but also in previously unmapped regions, such as an area west of Puerto Carreño between the Meta and Orinoco rivers in Colombia, South America. This project, incidentally, was begun and completed in the period between

December 1972 and February 1973, an accomplishment that could not have been made in any other way.

Monitoring the water supply in dry areas

In warm, arid regions with plenty of sunshine, copious crops can be grown with irrigation, but little or nothing of value can be grown without irrigation. Under such conditions, an accurate forecast of the amount of water that will be available for irrigation at critical times during the growing season is the most essential information that can be provided to a farmer. Some kinds of crops require more water than others, so the decision as to what crops to plant may be critically dependent on information about the expected water supply. Furthermore, unless he can be reasonably sure of being able to apply at least some minimum amount of water to a field at the proper time, the farmer will be better off not to plant at all. Similarly, the officials who manage irrigation districts must plan and contract for the distribution of available water in such a way as to maximize the production of needed crops.

The use of remote measurements of precipitation, stream flow and reservoir storage, telemetered by satellite, and the estimation of run-off from mountain snow fields by means of satellite pictures have already proved their ability to increase the accuracy and timeliness of such forecasts and to reduce their cost. Moreover, remote sensing from aircraft and satellites can improve the effectiveness of maintenance and operation of irrigation systems. Something like half of the water is typically lost between the dam head input to a canal system and the crops in the fields, most of it apparently by leakage from the canals and ditches. Such leaks can be quickly spotted in satellite images by the pattern of soil moisture and vegetation which they produce. Too much irrigation not only wastes valuable water, but also can cause the water level in the soil to rise so high that excessive evaporation occurs, increasing the salinity of the soil and eventually destroying its fertility. Thus the farmer may also need advice on proper timing and amount of water to use in irrigation. Information from satellite sensors can be used to formulate such advice as well as to detect early stages of increased salinity so that measures can be taken promptly to flush out existing salts and prevent further build up.

Supplemental irrigation by water pumped from underground sources is sometimes used in areas of marginal rainfall. As noted in the previous chapter, locations that are geologically favourable for underground water can frequently be identified in satellite images. Also, the extent to which rangeland is being converted to this kind of agriculture by private farmers and the amount of water they are withdrawing from known aquifers can be quickly and accurately estimated.

The detection of plant diseases and pests

Plant diseases and insect pests are perpetual enemies of the farmer and forester and sometimes cause losses of catastrophic proportions. Curiously enough, loss of plant vigour due to black stem rust of wheat or oats, late blight of potatoes, or root rot disease of navel oranges (to name but a few) can often be detected in near-infrared photographs from high-flying aircraft or in infrared images from satellites before it is apparent to a casual observer on the ground. This is because the spongy internal tissue of a healthy leaf, which is turgid—that is, distended by water—and full of air spaces, is an efficient reflector of radiant energy. The near-infrared wavelengths pass through the thin external tissue, which absorbs blue and red and reflects green from the visible spectrum, and are strongly reflected by the spongy tissue underneath. When a plant loses vigour its water balance is disturbed, the spongy tissue collapses, and almost immediately there is loss of leaf reflectivity in the near infrared. This change can occur long before there is any detectable change of reflectivity in the visible part of the spectrum, since it takes some time for a change to occur in the quantity or quality of chlorophyll in the external cells. Eventually the chlorophyll will be affected, of course, and then the plant will take on a characteristic yellowish, sickly appearance that is evident both to the local observer and in the visual bands of remote sensing devices. In most cases it is not so easy to determine the agent that is causing the loss of vigour from satellite imagery alone; however, once a particular infection is known to have begun, its geographical extent can be monitored by satellite so that protective measures can be organized efficiently. In some cases the agent can be identified by the characteristic time of year or weather conditions under which it appears.

Weather conditions can also be used to predict the spread of insects and disease. The desert locust, which invades the northern half of Africa and the Near East to India and Pakistan, caused losses totaling more than $42 million (US) during a nine-year period. The locusts tend to breed in certain desert areas following a rain and to migrate along paths determined by air currents along the Tropical Convergence Zone. The occurrence of precipitation over the breeding grounds and the subsequent wind patterns can be recognized in meteorological satellite images. Similarly, a disease called the black stem rust, which seriously affects wheat grown on the Ganges plain in India, moves in from the south in the form of wind-borne spores when suitable high-altitude winds occur. These winds are generally associated with a depression forming in the Bay of Bengal and travelling over southern India. Such a weather system during the critical period for infection of the wheat (November) is a fairly reliable indicator for forecasting wheat rust epidemics and can easily be identified in satellite cloud pictures.

When plant diseases or pests have been recognized and the areas of infestation delineated, control measures must be taken. Monitoring of the completeness and effectiveness of such measures is another way in which satellites can be of help to agriculture. Pink cotton bollworm in the Imperial Valley, California, for example, is controlled by carefully regulating the period during which all cotton-plant material (remaining after the harvest) is ploughed under. Supervision of the control programme requires accurate and timely determination of plough-down effectiveness. With ground data-collection methods this survey has typically required about 128 man-hours; completing the task on time has been a continuing problem. Such a survey was recently completed in 16 man-hours of *Landsat* image interpretation, with greater accuracy and reliability than could have been achieved previously. Average annual losses to agriculture caused by plant diseases and insect pests in the United States are reported to be about $7·5 thousand million; thus a small improvement in the effectiveness of the continuing war against these 'enemies' could easily pay for the entire United States earth-observation satellite programme.

Watching the forests by satellite

Forests are perhaps even more susceptible to disease and insect damage than field and orchard crops because they are rather inaccessible from the ground. Infestations can sometimes become well established before being detected. Also forests are subject to damage by fire. Estimates of annual loss caused by forest fires in the United States alone range from $300 million to $500 million, and the loss caused by pests is even larger. Losses of similar or greater magnitude are believed to be sustained in other parts of the world. In Honduras, Central America, for example, the southern pine beetle killed half of the pine timber on 2·5 million acres (approximately a million hectares) in just 1½ years.

The spread of forest diseases and insect pests can be observed by satellite using the same methods as for cultivated crops. Fires can easily be observed in satellite infrared images and can be detected by their smoke in other bands as well. The resolution of *Landsat* is adequate for this purpose, but since it images any given area only once in 18 days, it cannot be used for monitoring purposes. The highest-resolution sensors on meteorological satellites have been used for 'fire watching', but are not capable of delineating the boundary of a fire with sufficient precision for management of the fire-fighting effort, and in any case provide information at most only twice a day. Geostationary satellites can monitor fires continuously, unless obscured by clouds, but the resolution is not adequate. It would seem that television cameras and scanning infrared sensors on aircraft will continue to be needed for fighting forest fires, although satellites can be helpful in early detection.

Fire-prevention strategy in high-risk areas can be developed more effectively with the help of satellite data. In the San Francisco Bay region, unusually cold weather recently killed thousands of acres of eucalyptus trees in plantings interspersed with conifers, which were unaffected by the cold. Because of its high oil content and heavy ground litter, dead eucalyptus represents an extremely flammable fuel. The fire risk has always been high in this area and the economic impact of fires is greater than in most other forested areas because of the interspersion of suburban dwellings. *Landsat* image evaluations are being used to locate the most critical areas, where aerial photography will be used in developing a fire prevention and control strategy and for rehabilitation of the forest. *Landsat* images have also been used to survey large burned-over forest areas for purposes of damage assessment and rehabilitation planning.

An aid to agricultural planning

The most valuable role that satellites can play in agriculture is as a source of data for making and revising estimates of total area planted in specific crops and of crop condition or probable yield. Such estimates are basic to agricultural programme planning and administration. They help farmers plan their future plantings and market strategy, they serve as a direct measure of land utilization and as prime indicators of demand for farm supplies and labour, and they provide the basis for planning in food processing and marketing industries as well as in a large segment of transportation. They can even be important in international relations, as experience in recent years has increasingly demonstrated. By using operator-controlled computer analyses of digital data from the *Landsat* multispectral scanner, taken on successive passes throughout the growing season, it is possible to identify separate areas planted in almost every commercial crop, to measure the planted area for each crop to an accuracy consistent with the size of the *Landsat* resolution element, and to determine whether there are any abnormalities in crop vigour or development calendar that should be considered in estimating probable yield. Taking into account both spectral signature (that is, relative intensity measured in different spectral bands) and change of this signature with time, early experimenters have been able to identify and separate a wide variety of crops with a reliability of 70–90 per cent. For example, ordinary field corn (maize) can be separated from popcorn by the difference in their tasselling time, which produces a sharp change in image characteristics. Weather conditions at particular periods in the growth cycle, such as the tasselling of corn or the heading of wheat, barley, or rye, can have an especially important effect on yield and quality of the crop. On a different time-scale, cultivated forests or pastures can be considered as 'crops' and surveyed in the same way.

Natural vegetation in forests and open rangeland is more difficult to classify because it usually consists of a mixture of species. Nevertheless, studies have been made which show the possibility of dividing natural vegetation into a variety of 'habitats' with characteristic kinds and sizes of plants or trees on each. These different habitats can be mapped using *Landsat* images, as was recently done for the states of Minas Gerais and Esperito Santo in central eastern Brazil. Managed rangeland, which falls somewhere in the gap between cultivated crops and natural vegetation, is especially suitable for satellite monitoring. The condition of the grasses, whether dry or lush and green, overgrazed or suitable for further grazing, can easily be determined. This kind of information has been used with great economic benefit to determine how best to move livestock from one grazing area to another for optimum production on the available land in a given region. It can also be used to determine how much work is being done to rehabilitate the rangeland. With the use of *Landsat* images, an updated assessment of rangeland seeding for the entire state of Nevada, was completed with just three man-days of interpretation time. Although reasonably good statistics exist for rangeland improvement activity on public lands, they are difficult to extract and summarize from the records of the many different agencies involved. It is almost impossible to obtain an accurate assessment of such activity on privately held lands.

The importance of satellite imagery in all of these agricultural and forestry surveys lies in the relative speed and economy with which a trained interpreter, knowledgeable in the area and equipped with a small amount of ground reference data, can produce useful information. An example can be found in the preparation of a vegetational resources map for the Feather River watershed in California. This watershed is critical to the economy of the state. Maintenance of vegetation cover and careful management of its use are of great importance. Vegetational resource maps have been compiled at different times from conventional black-and-white aerial photography, from high-flight colour–infrared photography, and from *Landsat* false-colour imagery. The latter two are more accurate and show more detail. The costs for the three methods of preparing the maps are in the ratio of 20:9:1. Obviously the cost of making and launching the satellite is not included; however, there are so many different uses for *Landsat* images that the *pro rata* satellite cost for each of them becomes very small.

Satellites look at people

Although men, women, and children are not ordinarily thought of as 'fauna', except perhaps in a biological sense, the applications of satellite technology in demography and anthropology are closely related to other subjects covered in this chapter.

Locating centres of population

During this era of concern about the limits of growth, reliable population statistics assume a greater importance than ever before. The obvious way to obtain such statistics is by the arduous and expensive process known as census-taking, which most countries carry out periodically. Even in 'well-developed' countries, however, intervals between one census and the next are rather long, and the accuracy of the results is far from perfect. In some of the less-developed countries census-taking is at best haphazard and sporadic. A significant part of the world population has never been counted—only estimated. For reasons that have already been discussed, it does not seem likely that human beings, any more than other animals, will ever be counted individually in satellite images. In fact, humans would be even harder to detect than other animals because they stay under cover most of the time. Aggregates of human habitations called villages, towns, and cities can be observed in *Landsat* images, however, and from such observations estimates can be made of the associated populations. In particular, trends and movements of populations can be estimated during the period between official census results or in the absence of any other reliable statistics.

In a recent study in Brazil, 280 towns and cities, ranging in size from less than 600 inhabitants to 185 000, were identified in *Landsat* MSS Band 5 images, magnified to a scale of 1:500 000 (1 millimetre represents 500 metres). The area of each city or town was measured with a planimeter and the area data were correlated with population data from the 1970 Brazilian census. A reasonable correlation (index = 0·8) between area and population was found, the relationship being approximately linear at about 3000 persons per square kilometre. In a few instances, however, the population density was found to deviate markedly from the norm. Whether these deviations represent cultural or environmental differences that could be identified is not yet known. A larger sample would obviously be desirable, classified according to age, topographic considerations, mode of transportation, state of economic and industrial development, and so on.

Managing the coastlines of the world

The coastal areas of the world not only provide marshes for marine-life breeding and wildlife habitat, but also waterways for commerce and beaches and coves for recreational purposes. These latter characteristics promote economic development and increases in population which may damage or destroy the natural resource unless carefully managed. Land suitable for industry or residences is limited by the water boundary on one side and by the requirement to be reasonably close to the water on the other. Consequently coastal land is generally in short supply. This shortage often leads to draining and filling of marshland, which reduces the breeding area for marine resources, and to dredging and canalization, which affects salinity and flow of silt. Improperly treated effluents from industries and residential communities add to the other detrimental effects. Consequently there is conflict between the need for coastal land development and the need for preservation of natural conditions. Planning and management require information on which to base decisions. Some of this information can most easily be provided by remote sensors on high-flying aircraft and satellites.

In the past, land-use maps and inventories were made on the basis of black-and-white photographs from aircraft at about 3650 metres altitude. It was necessary to make mosaics of 300–400 photographs to cover a 4650 square kilometre coastal area comprising three coastal counties on the Mississippi Gulf Coast. This required much time and effort, and the results were less than adequate because the individual frames were taken under different conditions of illumination and suffered from significant geometric distortion away from the centre of the image. The change to high-altitude (18 200 metres) photography with colour–infrared film has been a great improvement. Only 15–20 frames are required to cover the same area, and many features can be observed in the colour–infrared images that are not noticeable in black and white. A typical land-use map developed in this manner identifies about 20 categories of land use and requires about two months to prepare, using six photo-interpreters having moderate training and skill. It might be expected, therefore, that imagery from satellites, at still higher altitudes, would offer even greater improvement. A single *Landsat* image more than covers the areas described, but the resolution is correspondingly poorer. However, good photointerpretation techniques applied to an enlarged section of a *Landsat* image of this area can still delineate about five or ten categories of land use, and it is believed that this capability can be improved with 'first-generation' images (that is, those directly derived from digital data) and by comparing several images made at different seasons of the year. Computer-aided classification methods, which are being developed for this application, will be needed to handle the large amount of data.

Proper management of coastal areas also requires frequent monitoring of the condition of the marshlands and coastal waters, to determine the impact of human activity on them. In general, information about the marshes is determined by identifying and mapping marsh vegetation. Coastal water conditions are determined by observations of surface temperature, salinity, turbidity, and chlorophyll content, all of which can be measured from satellites, although perhaps not with adequate resolution for this purpose. This is one application where increased resolution of the satellite images at the expense of area covered would be of value.

Satellites assist survival

In the part of West Africa that lies just south of the Sahara Desert, an unusually long period of less than normal rainfall has caused the loss of much of the livestock and starvation or near starvation for much of the human population. Ultimately climatologists may be able to explain why this drought has occurred, but the current need is to plan for relief, rehabilitation, and resettlement of the affected people. In order to do this intelligently and effectively, use must be made of what anthropologists call 'carrying capacity' methodology. Carrying capacity is the number of hectares required to support a given number of persons or animals, or both, at a given level of technology. All such methodologies or formulae require information on the amount of land which can be put into cultivation at a given level of agricultural technology and also on the population size and distribution.

Although human settlements and cultivations in this drought-stricken area are close to the limits of resolution of present *Landsat* imagery, villages more than 250 metres in diameter and cultivations or fields of 10 hectares or larger can be identified in images made at the right season of the year. Villages can usually, but not always, be seen best in the dry season, because shade trees that keep their foliage during this period give a response in Band 7, and thatched roofs, manure, and other debris respond in Band 5. Cultivations are most easily observed at the end of the brief rainy season. During 1973, a single village residential area and the surrounding domain were intensively studied by means of on-the-ground measurements and aerial photographs, and this work was related to *Landsat* images, in which this and about 85 other similar villages could be identified. At this time, villagers in the area were still practising subsistence agriculture, mainly with hand tools. Small livestock, sheep, and goats were surviving fairly well, but larger stock had suffered. The population at that time was 486. The theoretical carrying capacity, based on normal rainfall, would have been estimated at about 1000 people. Through data taken from the *Landsat* images, the information obtained at this one site can be extended to an entire culture area.

Satellites help with the overpopulation problem

In parts of India, Pakistan, and Bangladesh no satellite images, methods, or formulae are required to know that the population is already near the limits of carrying capacity. People are starving to death. Yet there seems to be no way to increase food production in these regions fast enough even to keep up with population growth; nor are there now any arable wilderness areas into which these people could be resettled in the vast numbers required. The only acceptable alternative is birth control. Medical research has provided reasonably effective and economical methods which, if widely used, could reduce the population growth to zero or below. The immediate problem, therefore, is to persuade hundreds of millions of poorly educated people that it would be good for them to use birth control and to teach them how to do it. Skillfully prepared television programmes, disseminated on a regular basis by means of broadcast satellites directly to village receivers, along with other programme material, may provide at least part of a solution to this very urgent problem. The technological capability is available, as described in Chapter 4.

It is very difficult to put a monetary value on the potential applications of satellite technology for anthropology, demography, or land-use studies, a few early examples of which have been described here. However, if satellites can indeed help the human race to keep its own numbers in reasonable balance with its resources and environment, and to manage its activities in such a way as to preserve that environment, the space programme will surely not have been in vain.

9 A laboratory for science; an observatory for astronomy

Astronomy compels the soul to look upwards.
Plato, *The Republic*, Book VII

In the previous five chapters I have been writing mainly about what could be referred to as the 'practical applications' of satellite technology. From the lofty viewpoint of the satellite, we have been looking downward. But in this chapter, consideration will be given to use of satellites as places for scientific work and as bases from which to make astronomical observations. The viewing direction will be to the immediate surroundings and upward rather than downward, and the focus will be on knowledge and understanding rather than on direct practical benefits.

Experiments without gravity

The three special aspects of the satellite environment, discussed in Chapter 3, that are difficult to reproduce on earth are a nearly perfect vacuum of almost infinite extent, the availability of heavy highly ionizing cosmic-ray particles, and the balancing of gravitational and dynamic acceleration to produce a condition that is equivalent for most purposes to the complete absence of gravity. The first two can be approximated closely enough for most experimental work in terrestrial facilities such as large vacuum chambers and ion accelerators. However, there is no way to reproduce the absence of gravitational acceleration on earth except in an aircraft flying a path that exactly matches a vacuum ballistic trajectory, or in a tall evacuated drop tower. In either case, the experiment can last only a few seconds, or tens of seconds at most. Any experiment which requires a sustained acceleration field less than that of gravity for more than about half a minute cannot be done in any conceivable terrestrial facility. In a satellite, however, the net acceleration field can be maintained at very nearly zero for as long as any experiment may require—be it hours, days, or years.

Many of the familiar physical processes, such as the burning of a candle, the boiling of water, and the pouring of a liquid from a bottle into a glass, depend on gravity. Without gravity there is mass but not weight. Hot gas or liquid ceases to rise, so there can be no such phenomenon as thermal convection. Mixing can only occur by the slow process of molecular diffusion or by forced convection, using a fan for example. The flame of a candle quickly goes out, 'drowned' in its own combustion products, although the wick may continue to glow for some time using the small amount of oxygen it can obtain by diffusion. Liquid will not pour out of a bottle; it is held inside, weakly, by adhesion and surface tension, although some of it can be shaken out like tomato ketchup out of a particularly recalcitrant bottle. Once outside the bottle, the liquid forms itself into spherical balls that float aimlessly around until they bump into a surface of some kind, whereupon they either wet it or splash away as smaller balls, depending on the nature of the surface and the liquid, and on the size of the ball.

Only miniscule forces are required to hold a sphere of liquid or a solid object, for that matter, stationary in a desired position, or to move it slowly about. Such forces can easily be applied by electrostatic or electromagnetic techniques. The object can be heated by electromagnetic induction or thermal radiation or cooled by allowing it to radiate to a cooler surface. Thus a material can be melted, combined with another material if desired, and cooled until it solidifies again, all in a vacuum or any controlled atmosphere, and without being touched by a container of any kind. The advantages of this procedure lie not only in avoiding any possibility of contamination from a container or crucible, but also in the capability to supercool some materials far below their normal temperature of solidification, because of the absence of any wall surfaces on which crystals can nucleate.

Basic aspects of crystal growth can better be observed in the absence of thermal convection currents in the melt or solution from which they are grown. Non-homogeneous mixtures or suspensions remain stable indefinitely and do not settle out. Sharp separations of different materials in solution or suspension—protein molecules, for example—can be achieved by electrophoresis in a liquid column rather than in a gel or other matrix, because convection currents are not superposed on the slower migration of the particles under the influence of electric forces. The design of high-speed rotating machines, such as centrifuges, can be simplified, because the bearings do not need to support the weight of the rotor and some vibrational modes are not excited.

All living organisms have evolved in, and are therefore adapted to, the earth's gravitational field. How has this influenced their evolution and what role does gravity play in their continued survival? Clearly such questions cannot be answered satisfactorily without long-term experiments conducted in a laboratory where gravity can be eliminated. Satellites, of course do make such laboratories possible.

Fundamental gravitational physics

Fundamental experimentation in the field of gravitational physics has been limited because of its extreme difficulty, yet it is much needed to provide support for one or another of the magnificent theoretical edifices that have been erected in recent years, beginning with Einstein's theory of general relativity.

One possible satellite experiment would consist of four precisely spherical gyroscopes in a special satellite that would be placed in an approximately circular polar equatorial orbit at about 800 kilometres altitude. Two of the gyroscopes would have their spin axes oriented parallel to the earth's rotational axis, and the other two perpendicular to it. According to Newtonian physics, the spin axes of all four gyroscopes should remain fixed with respect to distant stars. General relativity, on the other hand, predicts a precession of about 7 seconds of arc per year for the first pair and about 0·05 seconds of arc per year for the second pair. Other competing theories predict significantly different values. In order for this experiment to be of value, many severe technological problems will have to be solved, not the least of which is that of making a gyroscope with a drift rate due to extraneous torques on the order of 0·001 seconds of arc per year. This is orders of magnitude less than the best earthbound gyroscope, but it may be possible to achieve in a satellite, where the supporting forces are negligible.

If successful, this experiment would be the first to have the capability to distinguish between the gravitational field of a stationary earth and of a rotating earth. The fact that the earth rotates is, of course, not in question. However, an experimentally verified theory for the gravitational interaction of moving masses will be important in understanding phenomena such as gravitational waves, neutron stars, and black holes, which are on the frontier of modern astrophysics and cosmology.

Solar–terrestrial physics

The broad field of science called solar–terrestrial physics is the study of the particle, magnetic-field, and radiation environments of the sun and earth, including their interactions and the dynamic processes involved. Although such studies had their beginnings in terrestrial observations of the geomagnetic field, aurorae, airglow, and the reflection of radio waves from the ionosphere and in ground-based solar astronomy, they have been revolutionized through the use of instruments on rockets, satellites, and interplanetary spacecraft. Since Van Allen's experiment on the first satellite launched by the United States (see p. 29) many satellites instrumented for solar–terrestrial physics have been launched, some carrying as many as 25 different experiments, others only a few. A veritable flood of data and scientific papers has resulted.

During the past decade and a half, the main features of the geometrically trapped radiation have been fairly well described and explained, especially the part closer to the earth than 5 or 6 times the earth's radius and at latitudes lower than about 45°, where the lifetime of trapped particles is long and conditions are relatively stable. Advances have also been made in the description and understanding of the outer portion of the magnetosphere where the particle motion is still controlled primarily by the earth's magnetic field, of the interplanetary medium where extensions of the solar field are in control, and of the complicated interface between these regions.

Although it occurs infrequently here on earth, plasma—a gas consisting of charged atomic particles called ions—is the normal state of most of the matter in the Universe. Plasma physics, which deals with the interaction of this ionized gas with magnetic and electric fields, governs many, if not most, astrophysical phenomena. Plasmas can, of course, be created in the laboratory. Much work of this kind is associated with attempts to learn how to obtain useful power from nuclear fusion. However, the physical scale of such work is so small, and the conditions are so perturbed by the necessity to confine the plasma and to introduce probes for measurement purposes, that it seems impossible to attack the fundamental problems of cosmic plasmas in the laboratory. There is just no way to scale down the spatial extension, temperature, and density to equivalent laboratory conditions. Satellites, however, make it possible to use the astrophysical plasma 'laboratory' that already exists in nearby space, where a wide variety of plasma processes can be observed in action and where, on a few occasions, it has been possible to alter the local conditions so as to perform experiments in the classical sense. The controlled injection of plasma by means of a nuclear explosion (*Argus, Starfire*) is such an example.

Satellite instruments used for measuring particles and fields in the magnetosphere and the interplanetary medium include channeltron electron-multiplier detectors and electrostatic analysers, for the very-low-energy particles, and scintillator and solid-state semiconductor detectors for high-energy particles. Various absorber materials, magnetic fields, electronic pulse-height analysers, and coincidence circuits are used to measure charge, energy levels, and direction of arrival. Magnetometers include flux-gate instruments, for measuring steady-state and slowly varying fields, and search-coil devices for measuring rapid changes. Instruments for measuring steady and rapidly changing electric fields, sweep-frequency radio receivers for analysing the radio emission from 'blobs' of hot plasma remote from the satellite, and radio-frequency impedance-measuring devices to detect the effect of the local plasma surrounding a satellite antenna are other research tools that have been found useful.

Satellite orbits must be arranged to cover the entire region in which the earth's magnetic field interacts with the solar wind (plasma streaming outward from the sun at supersonic speeds), including the areas referred to as the bow shock, the magnetopause or transition region, and the magnetosphere itself. This region extends from about 20 times the earth's radius in the solar direction to beyond the moon's distance in the antisolar direction, and of course includes polar as well as equatorial directions. In order to be able to distinguish spatial from temporal variations and to measure the propagation velocity of plasma shock waves and other instabilities, clusters of two or three closely, but not too closely, spaced satellites are needed. Whenever possible, data from earth satellites are combined with similar data from interplanetary spacecraft orbiting the sun both inside and outside the earth's orbit.

The upper atmosphere

Another aspect of solar–terrestrial physics is the influence of the magnetosphere and of solar electromagnetic radiation on the upper atmosphere. Although it contains only a small fraction of the total mass of the atmosphere, the upper atmosphere constitutes a vast photochemical factory that absorbs a large amount of solar radiation, some of which would otherwise be harmful to land forms of life and all of which ultimately affects climate. It contains a plentiful supply of free electrons that reflect and sometimes absorb radio waves and make possible large electric-current systems that affect the earth's magnetic field. Even manufacturers of such mundane things as aerosol spray cans are now beginning to be concerned about the upper atmosphere because of scientific predictions that chlorine from the freon used as a propellant in such devices may accumulate there in quantities sufficient to change the photochemical balance and reduce the amount of ozone, which strongly absorbs harmful solar ultraviolet radiation. There are similar concerns about the exhaust from large rocket boosters and high-altitude supersonic aircraft.

The physics and chemistry of the upper atmosphere, collectively referred to as aeronomy, have been studied by means of ionospheric, auroral, and airglow measurements from the ground and by measurements of variations in the geomagnetic field at the earth's surface. Direct observation of temperature, density, and composition can be made using research aircraft up to about 20 kilometres and using balloons up to 30 or 40 kilometres. However, aeronomy also requires measurements of radiation from the solar disc at X-ray and ultraviolet wavelengths, of the charged particles (which precipitate into high-latitude regions and cause the aurorae) and of the magnetic field pattern at heights just above the atmosphere. This magnetic-field data, together with similar data from a network of ground observatories, can accurately define the current flow in the atmosphere. In addition, *in-situ* measurements of the ionic, atomic, and molecular composition of the atmosphere in the altitude region from 40 kilometres out to about 300 kilometres are required. Direct data from this region are relatively sparse because aircraft and balloons cannot fly so high and satellites normally cannot fly so low. It is therefore sometimes called the 'ignorosphere'! Sounding rockets can obtain data in this region, but such data can provide only a limited number of 'snapshots' of a small part of the situations, because each rocket can make measurements at only one location and only for a few minutes.

In order to help fill some of this gap, a special satellite designated as *Atmosphere Explorer C* (alternatively *Explorer 51*) was launched in December 1973. This satellite contains an orbit-adjust propulsion system capable of overcoming the effect of drag forces down to altitudes as low as 120 kilometres. Its experimental sensors include ultraviolet spectrophotometers, neutral, positive, and negative ion mass spectrometers, a neutral atmospheric temperature spectrometer, a low-energy electron spectrometer, a visual airglow photometer, a cylindrical electrostatic probe, a retarding potential analyser, and a sensitive three-axis accelerometer. By means of this satellite it has been possible to measure directly the constituents of the atmosphere down to 129 kilometres over the whole earth, including the very important auroral zones, and to study all of the photochemical reactions that occur. Also, it has been found that this region of the atmosphere is in a highly dynamic state resembling, on a much larger scale of course, a very stormy sea. Strong gravity waves, excited by heating in the auroral zones, have been observed and vertical motions of the neutral (not ionized) components as great as 100 m/s have been measured. Data from *Atmospheric Explorer C* apply directly to the intricate problem of sorting out the interactions between the ionized and neutral parts of the atmosphere, on which scientists are working diligently. This is an especially interesting problem because it directly relates to the question of how solar disturbances eventually influence the weather down at the earth's surface.

Cosmic dust

One other area of science that is being pursued by means of satellites is the study of micrometeoroids or cosmic dust. Meteoroids are the smallest pieces of solid matter that exist in the solar system. They vary considerably in composition and appear to travel in orbits around the sun. Some are captured by the gravitational field of a planet and sooner or later plunge into its atmosphere, where they become incandescent, and all except the largest are vaporized. Those that enter the earth's atmosphere and are visible as streaks of light in the sky are called meteors or 'shooting stars'. Occasionally a fragment of one of the larger meteoroids survives its fiery passage through the atmosphere and falls to the surface where, with a bit of luck, it can be recovered and studied in the laboratory. Such a fragment is called a meteorite. Meteoroid masses range from hundreds of kilograms to less than a picogram (one thousandth of a millionth of a gram). The number density increases as a steep inverse function of size (see p. 34). The term micrometeoroid refers to a little meteoroid, usually meaning less than about a milligram in mass.

Aside from the obvious spacecraft design problems posed by these little bullets, there is considerable scientific interest in micrometeoroids. Hundreds of thousands of tons of meteoric material, most of it in the form of microscopic dust, enter the earth's atmosphere each year, adding metallic ions to the upper atmosphere and probably acting as nuclei for the tiny ice crystals in noctilucent clouds which occur at altitudes of 80–100 kilometres and are observed during twilight in far northern and southern latitudes. The origin of micrometeoroids is not known. They may represent original planetesimals, left over after formation of the solar system. They may be fragments of asteroids, ground into dust by repeated collisions. They may be debris from comets, sloughed off after the evaporation of adhesive ices. Possibly all of these sources have contributed in some measure. In any case, whatever can be learned about micrometeoroids will contribute to a better understanding of the composition and early history of the solar system.

Radar has been used to study ionized trails in the atmosphere left by meteors too small to be photographed. Satellites and interplanetary spacecraft, however, can detect micrometeoroid encounters by a variety of techniques, including sensitive large-area microphones, arrays of pressure cells with thin walls that can easily be punctured, thin opaque membranes which, when punctured, allow sunlight to illuminate a photometer on the inside, and others. In the most sophisticated of these instruments, the meteoroid impinges on a tungsten grid target and vaporizes, producing a tiny puff of ionized gas which then passes through a time-of-flight mass spectrometer. By this and various other diagnostic techniques, the atomic composition as well as the mass and velocity of the meteoroid can be determined. Optical telescopes mounted in satellites or spacecraft are also used for meteoroid research in two ways. In one method, two or more telescopes, with offset but overlapping fields of view, register the passage of a meteoroid near the spacecraft by means of reflected sunlight. In the other, one or more slowly scanning telescopes measure the intensity distribution, colour, and polarization of zodiacal light, which is known to be sunlight, scattered or reflected at least in part by many small cosmic dust particles at great distances. Zodiacal light can be observed from the earth's surface at night under good seeing conditions, but it is so dim compared with the background of airglow and scattered light from the atmosphere that such observations are not very useful. A small telescope in a satellite, if properly designed to prevent stray sunlight from entering its aperture, is much more satisfactory. Such an instrument can also be carried on an interplanetary spacecraft, both closer to the sun and farther from the sun than the earth's orbit, thus obtaining data from which the distribution of the dust throughout the solar system as well as the total amount can be determined.

The development of astronomy and the reason for using satellites

The foundations of astronomy go back to a time when men had only their eyes, their brains, and a few sticks and stones to work with. The invention of the telescope started a revolution in this 'mother of the sciences' that continued with the application of photographic plates, precision measuring machines, spectrographs, scanning photometers, light amplifiers, precision mechanical mounts, and pointing controls—an almost bewildering array of technological devices. Telescopes have been made larger in order to gather more light from incredibly distant galaxies and more precise to achieve the best possible resolving power. Special telescopes with large fields of view have been developed for star mapping and others with special optics and filters to study the sun. Wavelengths have been pushed out to atmospheric 'window' areas in the infrared part of the spectrum, and the earth has been combed to find good mountain-top observatory sites where the air is thin and dry, the clouds infrequent, and the 'seeing' good.

Following the discovery of non-terrestrial sources of radio waves that could penetrate the atmosphere and be detected on earth, a new branch of astronomy was born; it was called radio astronomy. In its own way the radio telescope has been an even more revolutionary development than the optical telescope, for it has brought about the recognition of such intriguing phenomena as quasars, apparently the most distant and powerful astronomical objects yet discovered, pulsars, which are now thought to be small but incredibly dense stars consisting mostly of neutrons, gigantic clouds of cold cosmic gas in which a variety of elementary chemical compounds have formed, and a pervasive background of cosmic radio waves that may be the remnant of the colossal burst of energy in which the Universe was born. Radio astronomy has also provided information about processes in the solar corona that could not be detected at optical wavelengths, and it gave the first evidence of the magnetically trapped radiation around the planet Jupiter and the high surface temperature under the clouds of Venus. The radio telescope is the only means astronomers have to probe the crowded nucleus of our Galaxy. Now, with the addition of a transmitter, radio astronomy has become radar astronomy, which has revealed some of the surface features of Venus and provided a reliable measurement of the periods of rotation of Venus and Mercury.

Considering the advances in astronomy that have resulted from opening up some of the radio-wavelength part of the spectrum, it is not hard to understand why astronomers and astrophysicists have been eager to extend their observational capability to other wavelengths. In fact all parts of the spectrum from the longest radio waves to the shortest (most energetic) gamma-rays carry significant information about physical processes in the Universe, and the information at different wavelengths is often complementary, that is, it can be combined to achieve a better understanding of what is going on than can be obtained from any limited portion of the spectrum alone. Earthbound instruments, however, can make observations only at radio

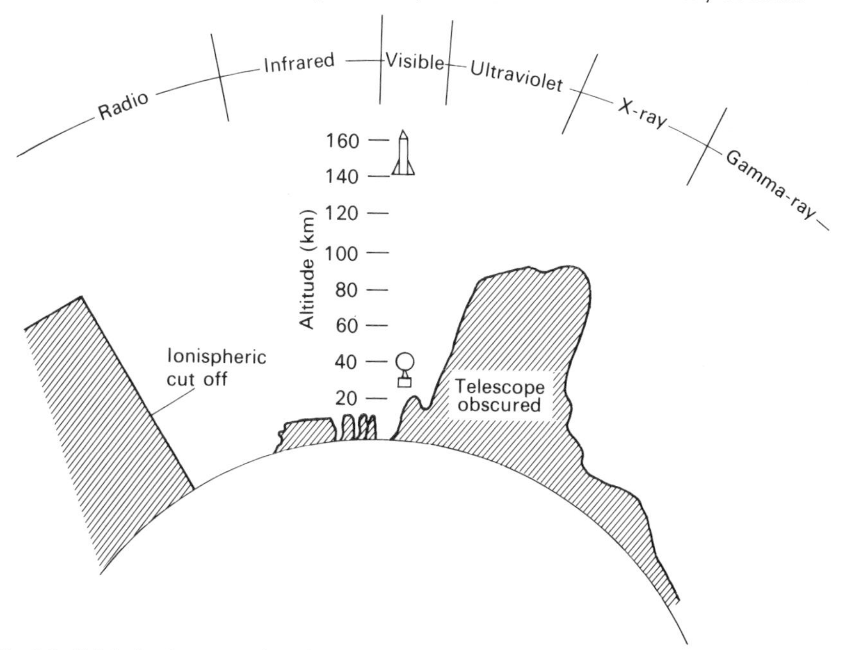

Fig. 9.1. Height in the atmosphere below which observations cannot be made in different parts of the electromagnetic wave spectrum.

wavelengths between about 10 metres and 1 centimetre, at a few infrared wavelengths, and at visible wavelengths from violet to near infrared (approximately 0·3–1·0 micrometres), a very small part of the total electromegnatic spectrum. This limitation is not imposed by inadequacy of instrument or telescope technology, but rather by the inherent transmission characteristics of the atmosphere through which the instruments must 'look'. There is only one way to overcome such a limitation and that is to raise the instruments above the atmosphere, or at least above most of it. Fig. 9.1 shows qualitatively the height at which atmospheric transmission becomes tolerable for astronomy at different wavelengths. Large balloons and research aircraft can reach altitudes suitable for infrared and gamma-ray astronomy; however, they do have limitations in pointing accuracy, payload size, and flight duration. In the ultraviolet and X-ray regions there is no alternative but to use rockets or satellites. Rockets are limited to only a few minutes of observations on each flight. They have been very useful for pioneering observations in opening up new areas of astronomy, but as more is learned and more questions are asked, satellites must be used to get the answers.

Ultraviolet astronomy

Ultraviolet telescopes of good quality have been flown in two American satellites, launched in 1968 and 1972. The first of these orbiting astronomical observatories contained two complementary groups of experiments. One, designed by the Smithsonian Astrophysical Observatory, consisted of four 30-centimetre ultraviolet telescopes having different spectral sensitivity, each forming an image on a special ultraviolet-sensitive television tube with a 2° square field of view. During its planned 16 months of operation this experiment mapped one-sixth of the entire sky in four different ultraviolet spectral intervals, obtaining 13 646 observations of 5068 stars! Based on these data, a catalogue of ultraviolet bright stars has been published that will serve as a basis for much future work.

The other experiment package, by the University of Wisconsin, was not intended to produce images but rather to obtain precise ultraviolet luminosity and spectral data for a large number of sources, ranging from the earth's outer atmosphere and other planets in the solar system to extragalactic nebulae. It used several separate telescopes of 20-centimetre to 40-centimetre aperture with filter photometry and scanning spectrophotometry. Observations were made at a rate of about 12 per day over most of the four-year life of the satellite. Among the significant results is new data about the modification and scattering of starlight in the ultraviolet as it passes through the interplanetary medium. This data indicates the presence of small (0·01 micrometres radius) spherical graphite grains distributed through our Galaxy, that are similar to grains that are believed to be formed and ejected from the outer parts of old, cool, carbon-rich stars. Also, the light from other nearby galaxies was found to show an extreme excess of ultraviolet radiation which, it seems, can best be explained by the presence of similar particles in those galaxies. An unexpected result from within the solar system was the discovery of a large cloud of hydrogen around the comets Tago–Sato–Kosaka and Bennett. The size of these clouds was half the size of the sun! The presence of this hydrogen, along with hydroxyl and atomic oxygen, clearly indicates that a major constituent of the cometary nucleus is water in the form of ice.

The second orbiting astronomical observatory, called *Copernicus*, carries an 80-centimetre telescope with an echelle-grating photoelectric spectrometer, created by

Fig. 9.2. Astronomical Netherlands satellite (ANS).

the Princeton University Observatory. The pointing accuracy is very precise and the spectral resolution is 0·005–0·01 nanometres (a nanometre is equal to 0·001 micrometres). This resolution is fine enough to detect even intrinsically faint absorption lines of the interstellar medium. The ultraviolet region, for which the *Copernicus* telescope was designed, includes the principal absorption lines of many interesting atomic and molecular species. Early results show that deuterium (a heavy-hydrogen isotope) is more abundant than expected, but that heavy atoms such as carbon, oxygen, nitrogen, and phosphorus are underabundant in the denser interstellar clouds. Perhaps the latter are removed by selective condensation on interstellar particles.

By now, several other ultraviolet telescopes have been launched, including the Astronomical Netherlands Satellite (ANS), a co-operative Netherlands–USA project. The ANS, launched in August 1974, contains a 20-centimetre ultraviolet instrument and also two telescopes for X-ray observations. An artist's conception of this satellite is shown in Fig. 9.2.

X-ray and gamma-ray astronomy

As the wavelength at which astronomical measurements are made becomes shorter and shorter, the information contained generally relates to physical processes involving higher and higher energy. The thermal radiation of a black body at what would be called ordinary temperature, say 100–1000 K, has a maximum in the infrared. If the temperature is raised to 5000 K, the maximum radiation occurs in the visible part of the spectrum, or, if it is raised to 30 000 K, in the ultraviolet part of the spectrum. At temperatures of around a million degrees and higher (which, of course, can only occur in plasmas), thermal radiation peaks in the X-ray region. Emission in discrete lines or bands at radio and infrared wavelengths is produced by molecular processes, at visible, ultraviolet, and X-ray wavelengths by interactions involving electrons at increasingly energetic levels of the atom, and at gamma-ray wavelengths by high-energy nuclear interactions. The wavelength of synchrotron radiation decreases as the electron energy and the strength of the magnetic field are increased. In fact, no matter how produced, radiation is always emitted or absorbed in discrete packages or wavetrains called 'photons', each containing a small but precise amount of energy that is inversely proportional to the wavelength of the radiation. Because the photon energy is so small at the radio end of the spectrum (of the order of 10^{-24} joules for centimetre waves) such radiation is always described in terms of wavelength. At the X-ray and gamma-ray end, however, it becomes more convenient to describe the radiation in terms of photon energy, measured in electronvolts. An electronvolt is equal to the energy acquired by an electron when it is accelerated through a potential different of a volt (see p. 30). For example, at an X-ray wavelength of one-thousandth of a micrometre (one nanometre), the photon energy is 1240 electronvolts, or a little more than a kiloelectronvolt (keV). Photon energy can be increased, as from infrared to X-ray levels, by interaction of the photons with energetic particles, a process known as inverse Compton scattering.

Even the most highly polished mirror will not reflect X-rays very well at normal incidence angles. Therefore, ordinary reflector telescopes cannot be used. For the softer X-rays (250 eV, for example) a grazing-incidence reflecting telescope can be used; in fact one is being used in the Astronomical Netherlands Satellite (ANS). The mirror in such a telescope is shaped like the outermost part of a long, drawn-out, short-focal-length paraboloid of revolution. The inside surface of a bullet, with the nose missing, might be another way to describe it. Radiation which enters the large end or aperture is reflected by the walls at very low incidence angles and thereby concentrated on a detector at the focus. The detector is typically a gas-filled proportional counter with a thin plastic-covered opening. Other types of X-ray instruments include large-area proportional counters with the spectral sensitivity range controlled by the pressure and type of gas contained, and the thickness and material of the cover. A collimator can be used to provide directionality and an active anti-coincidence shield to discriminate against most energetic particles. At higher energies, crystal scintillation counters may be used, and at still higher energies, in the gamma-ray region, large spark chambers are probably the most useful. If sufficient intensity is available, as when observing solar X-rays, for example, or at the focus of a large grazing-incidence telescope, spectral lines in the X-ray region can be identified by Bragg crystal diffraction spectroscopy.

Satellites for X-ray and gamma-ray astronomy

Until very recently, X-ray and gamma-ray astronomical instruments were carried aloft almost exclusively by rockets and high-altitude balloons. Much good scientific work has been and will continue to be done in that way. A few were included as individual experiments in the payload of satellites intended primarily for other work. In 1970, however, the first United States small astronomical satellite, SAS-1, later named *Uhuru*, was launched from the Italian San Marco platform off the coast of Kenya. Devoted entirely to X-ray astronomy, it was instrumented with two sets of proportional counter banks and was used to survey the sky for new sources, to search for temporal variations in known sources, and to measure spectral distribution in the range from 1 keV to 20 keV. During its first two years of operation, the number of known sources was increased from about 35 to more than 200, and a wealth of detailed data was obtained. In 1972, SAS-2, instrumented primarily for gamma-ray astronomy, joined *Uhuru* in the sky, and in the same year an X-ray instrument built and operated by the University of London was flown on the orbiting astronomical observatory OAO-3. The latter can determine the position of discrete sources more accurately than *Uhuru*.

The Crab Nebula, quasars, and black holes

The era of X-ray astronomy has been in existence scarcely more than a decade but has already brought much new information about the most energetic phenomena of this violent Universe. Observations to date have revealed many X-ray sources that radiate far more power in the X-ray part of the spectrum than in the optical or radio parts. Some star-like sources in our own Galaxy have X-ray luminosities as great as 10^{30} (the number 1 followed by 30 zeros) joules per second, 1000 times as great as the total power radiated by our own sun at all wavelengths combined. Certain kinds of variable radio galaxies may reach 10^{36} joules per second, which is comparable to the entire energy output of the Milky Way.

About two-thirds of the X-ray sources catalogued to date lie close to the Galactic plane and presumably belong to our Galaxy. Prominent among these are a number of supernova remnants such as the Crab Nebula. In their gross emission characteristics, these sources are relatively stable. Other discrete sources in the Galaxy show high variability, from a random flickering to the extremely precise periodicity of pulsars and eclipsing binaries. A few nova-like X-ray sources have been observed, which flashed to a brightness greater than any other X-ray object in our Galaxy and then faded away in a few months.

The Crab Nebula, which undoubtedly resulted from a supernova explosion in A.D. 1054, is one of the most intensively studied objects in our Galaxy. It seems to be an enormous mass of gas with tangled filaments of magnetic field distributed throughout. Relativistic electrons (electrons moving at velocities close to the speed of light) spiral along these magnetic field lines to produce synchroton radiation ranging in wavelength from radio to X-rays. A pulsar contained within the nebula is almost certainly the collapsed core of the exploded star. Its pulses can be observed at radio, optical, X-ray, and gamma-ray wavelengths, the X-rays being by far the most powerful. A picture of the Crab Nebula, showing the approximate location of the pulsar within it, can be seen in Fig. 9.3. All of the evidence fits the model of a spinning neutron star with a dipole magnetic field of about one hundred million tesla! It is believed to have a radius of only about 10 kilometres but a core density of 3×10^{17} kg/m^3 and to rotate at about 30 revolutions per second. Its kinetic energy is being slowly dissipated by the production of relativistic particles, which in turn produce synchrotron radiation with a power of 10 000 times the solar luminosity, and this process continues more than 900 years after the event that gave birth

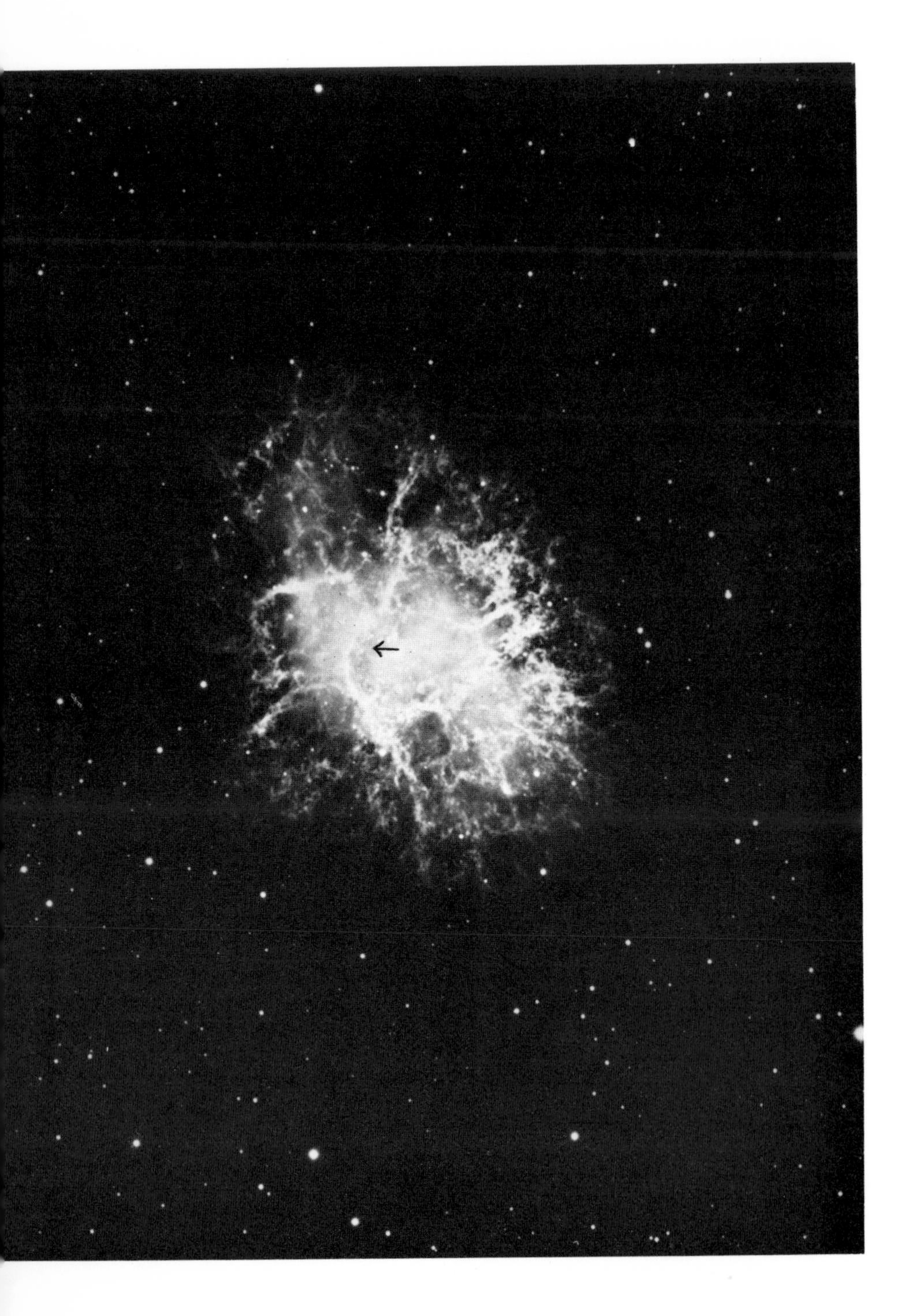

Fig. 9.3. The Crab Nebula. (Arrow points to approximate location of pulsar NP0532.)

to the pulsar! The loss of kinetic energy can be confirmed by the small but precisely measurable slowing down of the pulsar rate. Never before have scientists had such a clear and comprehensive view of so cosmic an event.

Not many X-ray sources are so well explained, however. Some appear to be binary stars, the radiation being produced by accretion of gas on to a compact, high-density component which may be a white dwarf or a neutron star. In at least one source (Cygnus X-1) the visible member of the binary is a blue giant star and the compact member is thought to be a black hole—that is, an object even more highly compressed than a neutron star. Such an object would have a gravitational field so strong that, according to the theory of relativity, no radiation, particles, or information of any kind could escape. Hence it could not be seen nor could it be detected in any way except by the influence of its external gravitational field. Theoreticians have predicted such objects; in fact they may be fairly common in the Universe. X-ray astronomy has helped to contribute the first physical evidence that seems to confirm their existence.

Looking beyond our Galaxy, astronomers have found galaxies in which violently explosive events seem to have occurred repetitively. Seyfert galaxies, which are characterized by small extremely bright nuclei, and quasars, which seem to be at vast cosmological distances from our Galaxy, are also interesting subjects for X-ray astronomy. Surprising as the intense visible and radio luminosity of quasars has proved to be, the X-ray and infrared emission can sometimes be orders of magnitude greater. How this vast amount of energy can be released strains the imagination of even the most ingenious astrophysicists.

From the viewpoint of cosmology, observations of the diffuse background of X-rays and gamma rays are highly important. The soft X-ray background can possibly be interpreted in terms of intergalactic gas, which may contain enough mass to determine whether the Universe is gravitationally 'closed', and will eventually fall back in on itself, or 'open', and will go on expanding forever. In the gamma-ray region, scientists would like to find a background of radiation that would indicate annihilation of matter by antimatter. Present theory indicates that equal amounts of each should have been created in the Universe, but so far even cosmic-ray astronomers have been unable to find any antimatter outside the particle accelerator laboratory. Are there perhaps separate galaxies consisting exclusively of antimatter as ours consists of matter? If so, there ought to be boundaries where the two kinds of galaxies interact, and these boundaries should be discernible by means of gamma-ray astronomy.

Optical astronomy

Although reasonably transparent in the visible part of the spectrum, the earth's atmosphere does interfere with ordinary astronomical observations, as any astronomer will testify. During the day, scattered sunlight makes the sky so bright that it obscures even the brightest stars. Only solar astronomy can be pursued in the daytime and even the solar corona, except for its brightest parts, is not visible against the bright sky. Until the advent of rocket and satellite technology, most of the corona could only be observed during the very brief and infrequent periods of total solar eclipse. At night there is a faint but bothersome background of natural airglow and of scattered light from man-made sources such as nearby cities and shopping centres. Sometimes, of course, there are clouds. Even on the clearest of nights at the best of sites there are atmospheric irregularities, small cells of slightly different temperature or humidity, that cause a telescope image to dance around or to spread out so that the effective resolution is less than that which should theoretically be obtained. This condition of poor 'seeing' varies from day to day, but never completely disappears. It is this seeing condition that limits the useful size of terrestrial telescopes. The 200-inch (5-metre) Hale telescope at Mount Palomar in California, has a theoretical image size of about 0·025 seconds of arc, but its characteristic seeing-limited images are 100 times larger—about 2·5 seconds of arc.

A telescope in space would be free of all of these disturbing effects and could produce images of exceptional quality 24 hours a day, every day. So far, no optical telescope has been flown on a satellite except for solar astronomy and for use at ultraviolet wavelengths. However, a long-range programme has been started which should lead to the orbiting of a high-precision three-metre telescope during the early 1980s. When trained on a point source, this telescope will produce a diffraction-limited image 0·05 seconds of arc in diameter. It should be able to discriminate between two objects 15 kilometres apart at a distance of sixty million kilometres—roughly the distance from earth to Mars. Because of the lower sky brightness in space and because the desired signal is concentrated into a smaller image, the proposed large space telescope should also be able to detect sources that have previously been too faint. It is estimated, for example, that the space telescope will be able to reach sources some 75 times fainter or 8·7 times more distant than can the Hale telescope. Thus it will represent almost as important a step beyond the best modern ground-based telescopes as that from the telescopes used by Sir William Herschel in the nineteenth century to those of the present day.

Radio astronomy

Although ground-based radio astronomy has made a revolutionary contribution to the science of atronomy, the radio-astronomy programme in space has been rather modest because the atmosphere is reasonably transparent over much of the radio-frequency spectrum and there has not seemed to be any very urgent need to go beyond it. At wavelengths longer than about 10–20 metres, however, the ionosphere reflects and absorbs radio waves to such an extent that this part of the spectrum has not been accessible to ground-based astronomy. As it has become evident that the radio power of our Galaxy, and of many other radio sources, continues to increase toward longer wavelengths right up to the limit imposed by the ionosphere, some interest has been aroused in the possibility of making observations beyond this spectral limit by means of rockets and satellites.

In order to receive a radio signal efficiently, the antenna length must be of the order of a quarter to half a wavelength. In order to achieve good directional characteristics a reflector or antenna array having much larger dimensions is required. For a wavelength of around 300 metres (a frequency of 1 megahertz) the typical dimensions of a long-wave radio-astronomy satellite must be at least 150 metres and preferably several times that. Fortunately such structures need not have any great strength, stiffness, or dimensional precision, so they can be formed of expandable or extendible members.

Explorers 38 and 49

The first American radio-astronomy *Explorer* satellite (*Explorer 38*), launched in July 1968, had four antennae, each 229 metres long when fully extended. These were arranged in the form of two 60° Vs, one extending downward toward earth, used to monitor long-wavelength radio emissions from the terrestrial environment, and the other extending upward to make observations of radio sources in space. A dipole antenna consisting of two 18·3-metre sections was also deployed in a horizontal direction; it was intended to monitor radio bursts from Jupiter, the sun, or other possible sources. These antennae were made of pre-stressed thin beryllium–copper strip, stored on reels in the satellite during launch. When deployed by slowly unrolling the reels, the strip formed itself into 1·27 centimetre-diameter tubes that were stiff enough to maintain the desired configuration in the absence of any significant acceleration. The satellite was gravity-gradient stabilized, and for this purpose also deployed a libration damper boom. Associated with the V antennas were three radiometers (carefully calibrated radio receivers) capable of measuring at wavelengths from 32·4 metres to 667 metres in nine steps. A similar radiometer was associated with the dipole antenna along with two burst receivers, one a 32-step receiver covering the range from 54·3 metres to 1500 metres and the other a sweep-frequency receiver covering the range from 76·3 metres to 1230 metres.

One of the surprises resulting from the operation of this satellite was the large number of solar flares recorded—sometimes as many as 100 per day. The strongest of these appeared to be long-wave extensions of flares observed on the ground but many weaker ones had no ground-based counterparts. Radiation from the earth's magnetosphere was found to be stronger than expected, and good spectral data for the Galaxy were obtained in different directions. Ten times as much radiation was found to be coming from the spiral arms as from the directions between them.

The second radio-astronomy *Explorer* (*Explorer 49*) was launched in June 1973 and placed in orbit around the moon. It is similar in design and function to *Explorer 38*, except that the radiometers in the new satellite can cover a wider range of wavelengths, from about 23 metres to about 15 000 metres. The lunar orbit has the advantage of positioning the radiometers further away from the noise generated in the earth's ionosphere and magnetosphere. In addition, lunar occultations are used to determine the positions of radio sources much more precisely than would otherwise be possible considering the inherently poor directional qualities of such a radio 'telescope'.

Up to the present there has been no satellite radio astronomy at the very-short-wavelength end of the radio spectrum, where water vapour and other molecular constituents of the atmosphere block most of the radiation. This interval from about 10 millimetres to about 50 micrometres, which covers the transition from radio to infrared measurement technology, has not been widely used for observations, partly at least because of the lack of suitable instruments. This lack is now being overcome, and it may be expected that satellites instrumented for observations in the millimetre-wave and far-infrared part of the spectrum will be constructed within a few years. Meanwhile, observations of this kind will probably be made using balloons and high-altitude research aircraft.

Radio interferometry using satellites

Although the atmosphere is transparent from 10- or 20-metre wavelengths down to about 1-centimetre wavelenghts, there is one possible application for satellite radio astronomy in this spectral range. The resolution of an ordinary radio telescope, even with a very large reflector, is not nearly as good as that of a small optical telescope. In order to achieve better resolution at radio wavelengths, a technique called radio interferometry has been used. Two or more radio telescopes, spaced far apart, record signals from the same source on wide-band recorders synchronized by very-high-precision atomic frequency standards. When these recordings are properly combined, a resolving power can be achieved that is comparable to that of a hypothetical telescope with an aperture equal to the distance between the individual telescopes used to make up the interferometer network. Thus, if the distance or baseline is about 5000 kilometres, distant sources as close together as 0·05 seconds of arc can be resolved at a wavelength of 1 metre. Observations made with this very-long-baseline interferometer technique have shown that quasars contain components that are not resolved even at the longest baseline achievable on earth. It has been proposed, therefore, to put a radio telescope on a satellite in an eccentric orbit that extends to great distances from the earth, and to use it in conjunction with an existing terrestrial radio telescope. The objective would be to construct, in effect, an aperture synthesis system that could 'map' quasars to a resolution of the order of 10^{-5} seconds of arc—equivalent to one light-year at the limit of the visible Universe.

The relevance of science and astronomy

It seems unlikely that Plato could have imagined how far 'upwards' astronomy would compel our souls to look. It may be argued that such grand questions as the origin, extent, and ultimate fate of the Universe, and the source of its incredibly vast energies, are not now relevant to people here on earth. Of course, history has shown that, in the past, when scientists have laboured to satisfy their curiosity about what at the time were seemingly irrelevant phenomena—such as the nature of electricity and magnetism, the structure of the atom and its nucleus, and the source of solar energy—the consequences have been very relevant in a most practical sense of the word. But beyond all possible considerations of practical benefit, the sciences, and especially astronomy, fulfil a burning desire of mankind to know and to understand. Perhaps this desire is the spark that makes man's soul divine.

10 Man in the satellite

How magnificent! I see the earth, forests, clouds . . .
Yuri Gagarin, remarks at the beginning of the flight of *Vostok I*

On 12 April 1961, a Soviet cosmonaut, Yuri Gagarin, circled the earth once in a satellite called *Vostok I* and was recovered, alive and well, the first man to experience flight outside the atmosphere. Less than a month later American astronaut Alan Shepard also ventured into space on a suborbital flight of the *Mercury* spacecraft, and in August this flight was repeated with Virgil Grissom as pilot. During that same month, however, cosmonaut Gherman Titov made a flight of 17 orbits (25·3 hours). In February 1962 astronaut John Glenn made a flight of three orbits that was repeated in May with Scott Carpenter at the controls. Thus the manned-spacecraft programme began—as a technological race between the Soviet Union and the United States, with the Soviet Union clearly in the lead. This programme has had its notable successes, any list of which must include the seven *Apollo* moon landings by the United States. It has also had its tragedies—such as the fiery deaths of astronauts Ed White, Virgil Grissom, and Roger Chaffee during a ground check-out test of an early *Apollo* spacecraft, the fatal crash landing of Vladimir Komarov, when the parachutes apparently became tangled during re-entry of the new *Soyuz I* satellite which he was testing for the first time, and the quick, quiet deaths of cosmonauts Georgi Dobrovolski, Vladislav Volkov, and Viktor Patsayev in 1971, when a faulty hatch mechanism caused sudden depressurization of the *Soyuz* spacecraft in which they were coming home after a record-breaking stay of 24 days in the *Salyut* space station.

The hazards of space

Of the many men and one woman who have ventured into space, however, none has been killed or injured by the effects of radiation, weightlessness, or meteoroids—the three hazards of space most feared at the beginning. The real hazards, it seems, have been unreliability of equipment and the fallibility of human beings. One of the greatest practical benefits from the space programme may well prove to be the advances that it has required in predicting and improving the reliability of devices, systems, and operations. The term 'man-rated' refers to the highest reliability engineers can achieve.

Why men in space?

Strange as it may seem, in view of the well-publicized challenge and drama of manned space flight, there have been many serious debates about its desirability. Large, expensive spacecraft are required to put human beings into orbit, so that they can work there effectively, and return home safely at the end of the mission. Not the least item on the expense sheet, by the way, is the extensive engineering and testing needed to 'man-rate' the giant boosters, the satellites, and everything on board. Is this expense really necessary? What can humans do in satellites that justifies the great expense and risk of having them there? Almost any satellite observation or experiment of which I am aware can be automated, or at least mechanized for remote control by a human on the ground, at a cost considerably less than that of providing a human operator in the satellite. The one exception—the one experiment which cannot be done without a man in the satellite—is, of course, an experiment on man himself. And there are reasons why such experiments should be made. As the distance from earth increases, remote control becomes more difficult because of the communication time lag. At lunar distances, remotely controlled rover vehicles must move very slowly, as Soviet tests with the *Lunakhod* vehicles have shown. At earth–Mars distances such vehicles would barely be able to creep, covering at most only a few tens of metres an hour. Thus, despite the high cost of astronauts, it may be necessary to send them to Mars, or at least into Mars' orbit, for detailed scientific exploration of that planet. But, if so, extensive preliminary medical testing will have to be done in earth orbit, before such journeys can be undertaken.

Man's outstanding quality, as compared with electronic and mechanical systems, is his extreme versatility. He cannot move as quickly, as precisely, or with as much strength as a machine, nor can he do arithmetic as fast or remember as many numbers as a computer, or see, hear, smell, taste, or feel as sensitively as appropriate instruments—but he can do all of these things reasonably well and, what is more, he can decide almost intuitively what to do and when and why. This quality of versatility is most useful when there are many different experiments to be done or observations to be made the outcome of which may be uncertain over wide limits beforehand. Human versatility can also be very useful when something goes wrong in a satellite. On several occasions during the *Apollo* programme, failures occurred in spacecraft subsystems or instruments which might have terminated or greatly reduced the value of the mission. Working in close communication with engineers and technicians on the ground, the astronauts were usually able to repair these failures or to work around them.

The cost of bringing the human occupants of a satellite back to earth is a large part of the total cost of a manned satellite mission. But if the mission is one that requires the recovery of physical samples or photographic film, the returning astronauts can bring such things back with them at little or no extra cost, thus compensating to some extent for the high cost of their return flights!

Skylab

The American *Skylab* programme is one example of successful use of men in a satellite. The *Skylab* satellite was made up of four modules, the largest of which, called the orbital workshop, was constructed by modifying a spare *Saturn* rocket tank section about 6·6 metres in diameter and 14·7 metres long. This module contained roomy quarters for a crew of three, equipment and food lockers, water tanks, waste-management equipment, laboratory work areas, airlocks for scientific experiments, and deployable solar cells and batteries for supplying power to everything but the solar telescope. Habitable volume in this module was more than 930 cubic metres. The other modules were an airlock module which made it possible for the astronauts to go outside in their space suits without depressurizing the whole spaceship, a multiple docking adapter with two docking ports for the ferry spacecraft, that brought up the astronaut crews, and the *Apollo* telescope mount, basically a solar astronomical observatory. The telescope module was the only one that was not habitable; its displays and controls were located in the multiple docking adapter and were operated from that position. All four modules were launched together as *Skylab 1* on 14 May 1973, with no astronauts on board.

During this launch a deployable micrometeroid shield, which was also intended to function as a sunshade, was torn away by aerodynamic forces. In the process, it permanently damaged one of the solar cell panels and fouled the other so that it could not be deployed automatically. Later investigation indicated that this accident was caused by improper design and construction of venting paths under the shield. At the time, however, the problems to be faced were lack of any electric power from the orbital-workshop solar-cell arrays and a rapidly rising temperature in that module. Fortunately, the *Apollo* telescope mount had its own electric power system with separate arrays of solar cells that did deploy properly, and could supply some power to the other modules. Using this power, the whole *Skylab* assembly could be reoriented by command from the ground, and in this way the temperature rise inside was kept low enough so that equipment was not damaged, although it would have been too hot for people to endure for very long.

After careful consideration it was decided to send the first crew of astronauts to try to rehabilitate the *Skylab* by rigging a makeshift sunshade, several of which had

Fig. 10.1. *Skylab 1* as the last crew departed on return trip home. In the foreground is the solar telescope module with its solar-cell panels deployed and the round white thermal shield for its instruments. The two makeshift sunshades erected by the astronauts can also be seen in the foreground.

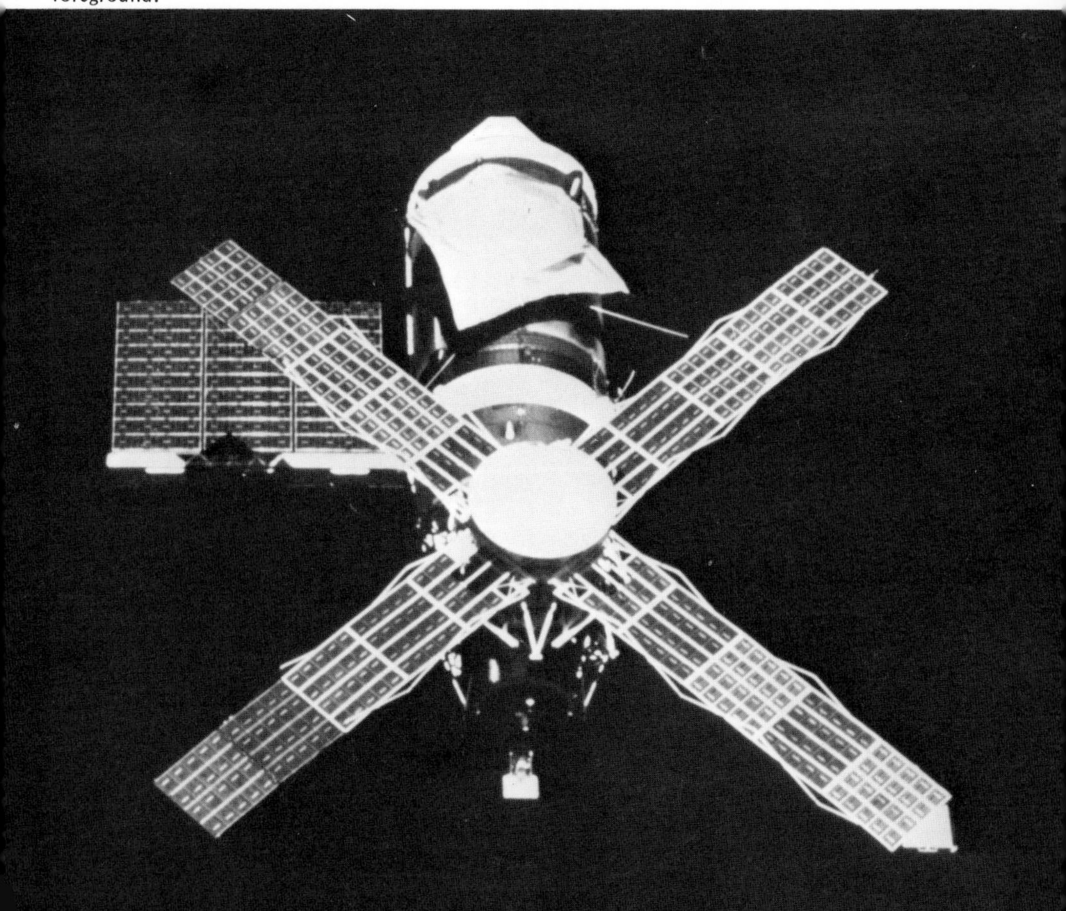

Fig. 10.2. Astronaut Jack Lousma engaged in extra-vehicular work during the second occupation of *Skylab*.

been hurriedly designed and constructed, and by manually deploying at least one of the solar-cell arrays, if that should prove to be possible. The crew took off on 25 May 1973 in *Skylab 2*, a ferry spacecraft similar to the *Apollo* command and service modules, but with additional propellant, additional water, and three 500 ampere-hour batteries. On the following day, the crew entered *Skylab 1* and successfully deployed a parasol-like thermal shield through one of the scientific-experiment air-locks. The temperature inside fell to a comfortable level shortly thereafter. On 7 June two of the astronauts climbed outside, cut away the debris of the meteoroid shield, and deployed the one undamaged solar panel, which then began producing electricity. *Skylab 1* was finally in business.

The first crew remained in orbit until 22 June, a total of 28 days, and completed about 90 per cent of their previously planned technical and scientific activities, despite the emergency repair work that had to be accomplished. A second crew was launched on 28 July 1973 and remained for 59 days, despite some early difficulties with 'stomach awareness'. These astronauts had to overcome a number of equipment problems, such as leaking thrusters in their ferry vehicle *Skylab 3*, faulty rate gyroscopes, a leak in the primary coolant system, and an inoperative regulator in one of the telescope-module power converters. They also erected a new and more effective sunshade outside the one that had been put in place by the first crew. Nevertheless, they accomplished more experimental work than originally planned, and returned safely and well. A third crew was launched in *Skylab 4* on 16 November 1973. This crew also did some minor repair work on *Skylab 1* and carried out a very extensive programme of observations and experiments. They returned on 8 February 1974 in excellent physical condition, after a stay of 84 days. This is several times longer than anyone had remained in orbit before *Skylab.*

Fig. 10.1 shows *Skylab* with its solar telescope and two make-shift thermal shields as it appeared during the final 'fly-around inspection', when the third crew began its journey back to earth. The three crews occupied *Skylab 1* for a total of about 171 days and worked outside in open space for more than 41 hours, doing repair work, installing instruments, changing film, cleaning telescope lenses and occulting discs, and a variety of other useful tasks. In Fig. 10.2 astronaut Jack Lousma of the second crew is shown working at one of these tasks. They brought back 182 842 frames of solar astronomy film, 40 200 frames of earth photography film, 72·5 kilometres of magnetic tape, and many samples produced during metallurgical and crystal-growing experiments. All this was in addition to a vast quantity of data and video displays telemetered to earth during the three missions.

Medical tests on the astronauts included monitoring their metabolic balance by carefully measuring and accounting for food and fluid intake, waste products, and energy expended. Normal physiological measurements were also made, such as changes in weight (not so easy in a weightless environment), body dimensions, body temperature, and blood pressure. Blood and urine samples were taken frequently and analysed. Motion-sensitivity tests were made in a rotating litter chair. Two critical tests of the cardiovascular system were regularly performed. One of these employed a device in which the lower limbs and torso could be subjected to reduced external pressure, thereby forcing the blood into the lower part of the body much as gravity does on earth. Response to this test is a measure of how well the astronaut's body will re-adapt to earth conditions upon his return. The other test involved monitoring the response of the cardiovascular system during sustained vigorous exercise on a bicycle ergometer, also used in the metabolic studies.

As expected, there was considerable variation in the results of these tests from one individual to another. All tended to lose some weight but not enough to be of concern. Most evidenced some weakening of cardiovascular response, but not to a serious extent. All suffered some decrease in number of red blood cells in the circulating blood-stream; this varied from about 5 per cent to about 20 per cent, depending on the individual, and may have been related more to the high oxygen content of the atmosphere used in the spacecraft than to weightlessness. The most encouraging result was that all of these effects tended to level off with time, indicating that reasonably stable conditions had been achieved. Regular daily exercise periods were found to be very desirable and were increased for the second and third crews. Because of the relatively low altitude of the *Skylab* orbit (about 430 kilometres circular at 50° inclination) none of the crews was subjected to any significant radiation dose.

The earth resources experiment package (EREP) on *Skylab* included high-resolution photography on conventional aerial and infrared–colour film and on black-and-white film in spectral bands corresponding to those of the *Landsat* scanner. It also included a multispectral scanner covering wavelengths from the blue end of the visible part of the spectrum to 12·5 micrometres in the thermal infrared region, an infrared spectrometer, an L-band microwave radiometer, and the S-193 microwave radiometer–scatterometer–altimeter mentioned on p. 114. These cameras and other sensors were all mounted on the outside of the multiple docking adapter, with the controls inside. For most earth observations it was necessary to manoeuvre the whole spacecraft away from its normal solar orientation to point the sensors along the local vertical. This constraint, together with the need for proper sun angle on the ground, crew sleep cycle, and other demands on the crew, limited the time that could be devoted to earth observations; however, at least 20 different EREP investigations were carried out, and *Skylab* data have been sent, along with *Landsat* data, to well over 100 users in many parts of the world. Microwave radar scatterometer and altimeter sensors had not previously been flown on a satellite. These were used in studies of the sea surface during conditions ranging from dead calm to full hurricane winds. The astronauts not only pointed the EREP instruments, changed the film, and made minor repairs, but also made visual observations and took pictures of special features such as volcanos, floods, and earthquake damage with hand-held cameras.

A multi-purpose electric furnace, also located in the multiple docking adapter, was used for materials-science experiments. Crystals were grown by controlled melting and solidification and by a vaporization–condensation cycle. The results tended to confirm previous expectations that larger, more perfect crystals could be grown in a space environment than on earth. Other experiments involved metal-sphere forming, cutting, welding, and brazing. All of this work is at a very early stage of development. For the most part, preliminary experiments could not usefully be made on the ground; therefore *Skylab* really represented the first chance to find out what would happen in the absence of gravity and what unexpected problems would be encountered. Much was learned.

One minor experiment worth mentioning was proposed by a young student. Its purpose was to find out whether a spider can weave its characteristically beautiful web without the pull of gravity to orient it. For this purpose two spiders, named Arabella and Anita, were carried on board, fed, and nurtured by the astronauts. At first they were rather confused, but eventually they did learn to spin webs of reasonable symmetry.

Important and interesting as all of the foregoing investigations have certainly proved to be, the area in which *Skylab* has contributed the most to science is that of solar astronomy. With 50–100 times better geometrical and spectral resolving power than ever before, the *Skylab* telescopic instruments probed the solar atmosphere from the photosphere to the far reaches of the corona, revealing many new details of the plasma interactions in this immense natural 'laboratory'. These instruments consisted of a white-light coronograph, an X-ray spectrographic telescope, an ultraviolet scanning spectroheliometer, an extreme ultraviolet and X-ray telescope, a coronal extreme ultraviolet spectroheliograph, and a chromospheric extreme ultraviolet spectrograph. Usually several of these would be used simultaneously. Supporting equipment enabled the astronauts to point the instruments precisely in the desired direction and provided position references on the face of the sun, so that satellite observations could be correlated with those made from the ground. All of the equipment worked well except for minor trouble with protective aperture doors and one of two television displays, which were repaired or replaced by the

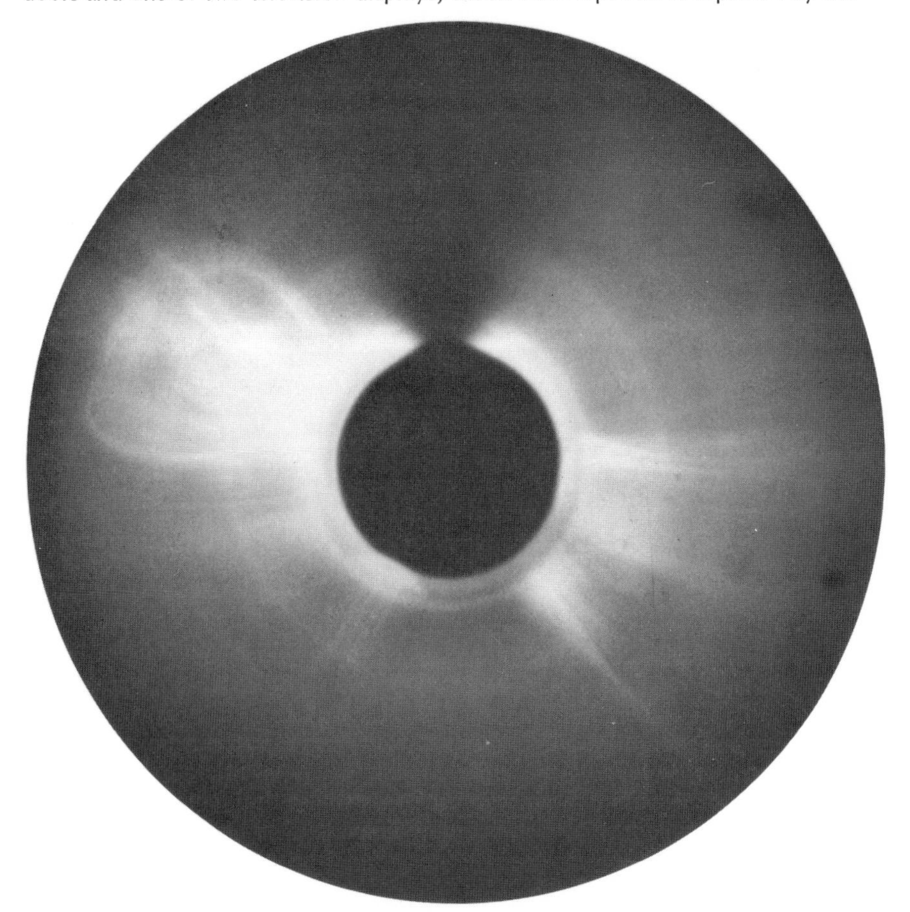

Fig. 10.3. Solar corona as photographed by white light coronograph on *Skylab*. Note strong coronal hole at top and enormous bubble of coronal gas being blown off by solar flare at upper left.

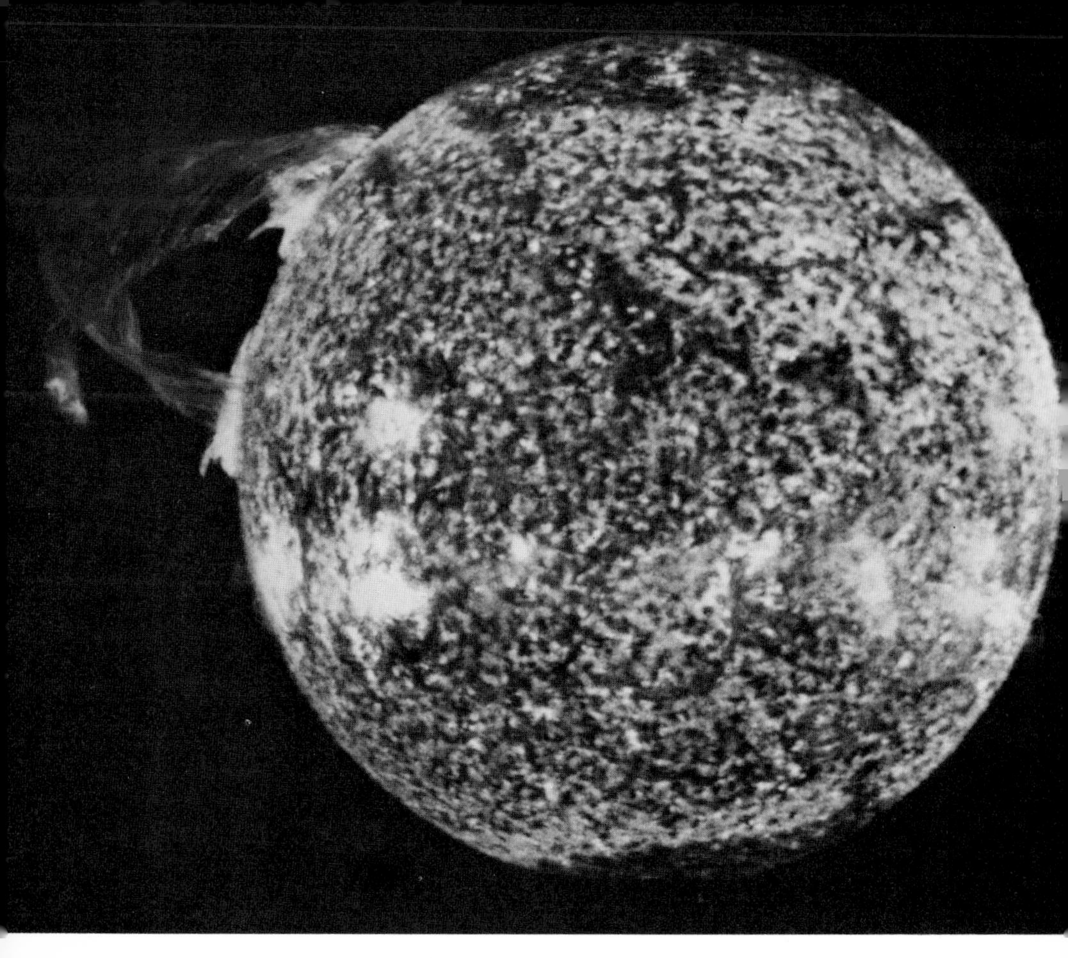

Fig. 10.4. Photograph of the sun at wavelength of ionized helium by the Naval Research Laboratory's extreme ultraviolet spectroheliograph instrument on *Skylab*. Very large solar prominence can be seen at the upper left. Supergranulation network that covers the entire disc is less prominent at poles.

astronauts. Some 27 specific observation programmes were defined and carried out during the three *Skylab* missions, 23 being specifically directed toward problems of solar physics and four to non-solar observations, such as that of the comet Kohoutek, which happened to come along at just the right time for the third crew.

Skylab photos have shown that essentially all of the solar corona is magnetically confined. Where the magnetic fields vanish or become essentially radial, coronal holes were observed—vast regions of abnormally low density and temperature through which the sun apparently loses at least some of the material that makes up the solar wind. Such holes are always present over the polar regions, but often occur at other latitudes as well. They persist over more than one solar rotation and do not show differential rotation (faster at the equator than at the poles) as do sunspots. A prominent coronal hole can be seen in the polar region at the top of the white light coronograph image shown in Fig. 10.3.

Many fine details of coronal form are revealed for the first time in *Skylab* films such as Fig. 10.4, made with the extreme ultraviolet spectroheliograph. Solar prominences can be seen as 'ropes' of twisted filaments, which unwind as a prominence erupts. While quiescent the core of a prominence remains relatively cool; as it becomes active it rapidly heats up and emission appears in the ultraviolet light of highly stripped ions. Perhaps magnetic induction that accompanies the unwinding of the magnetic field lines supplies the energy for this heating. Flares were found to occur most often at the feet of such loops, more rarely at the top of a loop, and sometimes simultaneously in both regions. When these explosive flares occur, great 'bubbles' of intensely hot gas hundreds of times larger than the earth are shot out into space, as shown in Fig. 10.3. When such bubbles reach the earth, they disturb the magnetosphere, causing magnetic storms and aurorae.

The corona also appears to be dotted with many bright, point-like X-ray and extreme ultraviolet 'micro-flares', which last only a short time. As many as 1500 of these X-ray bright points may appear in a day. It has been suggested that they are manifestations of the process by which the sun brings new magnetic flux up to the surface. The supergranulation network, prominently seen in Fig. 10.4, has been found to extend higher into the base of the corona than expected. Spicules, small fountain-like jets of gas that can be seen in visible light projecting above the solar limb like patches of tall grass, are even more prominent in the extreme ultraviolet. It is thought that they may create acoustic (hydromagnetic) waves that heat the corona to its multimillion-degree temperature.

The wealth of information obtained from *Skylab* after its inauspicious beginning has demonstrated, even to sceptics, that man can be very useful and perhaps even cost-effective in large complex missions such as this. It will take a few years for the scientists and engineers to digest the *Skylab* data and to plan for the next steps. Therefore *Skylab*, as such, will not be repeated, although *Skylab 1* will remain in orbit for a number of years and can be revisited by future astronauts if desired. Instead, the final scene in the first act of this drama has been a joint United States–Soviet Union manned space mission known, as the *Apollo–Soyuz* Test Program (ASTP), in which an *Apollo* spacecraft launched from the United States made a rendezvous and docking in earth orbit with a *Soyuz* spacecraft from the Soviet Union.

Apollo–Soyuz

The *Apollo* spacecraft was one of the last two remaining from the manned lunar exploration programme, after it was terminated for budgetary reasons in early 1973. The *Soyuz* was a more-or-less standard Soviet manned earth-orbiting spacecraft (see Fig. 10.5). That which was new was the ASTP docking module, designed conceptually by both countries, but constructed in the United States for NASA. Its main function, of course, was to provide a mechanical coupling between the two spacecraft that would mate with the *Apollo* docking port on one end and the *Soyuz* docking port on the other and hold the two 'foreign' spacecraft firmly together as a single unit. In addition, it had to provide a transition for the astronauts and cosmonauts between the 5 p.s.i. (34·4 kilopascals) pure oxygen atmosphere in *Apollo* and the normally 14·7 p.s.i. (101 kilopascals) nitrogen–oxygen atmosphere in *Soyuz*, reduced during this test to 10 p.s.i. (68·8 kilopascals). Therefore, the docking module was equipped with a nitrogen supply, to dilute the pure oxygen from *Apollo*, and means for raising and lowering its internal pressure gradually, so that men could pass through in either direction with adequate time to adapt to the different environment at the other end.

The first docking was made on 17 July 1975. Two days later the two spacecraft were separated and the undocked *Apollo* was used as an occulting shield to block out the direct rays from the disc of the sun so that photographs of the solar corona

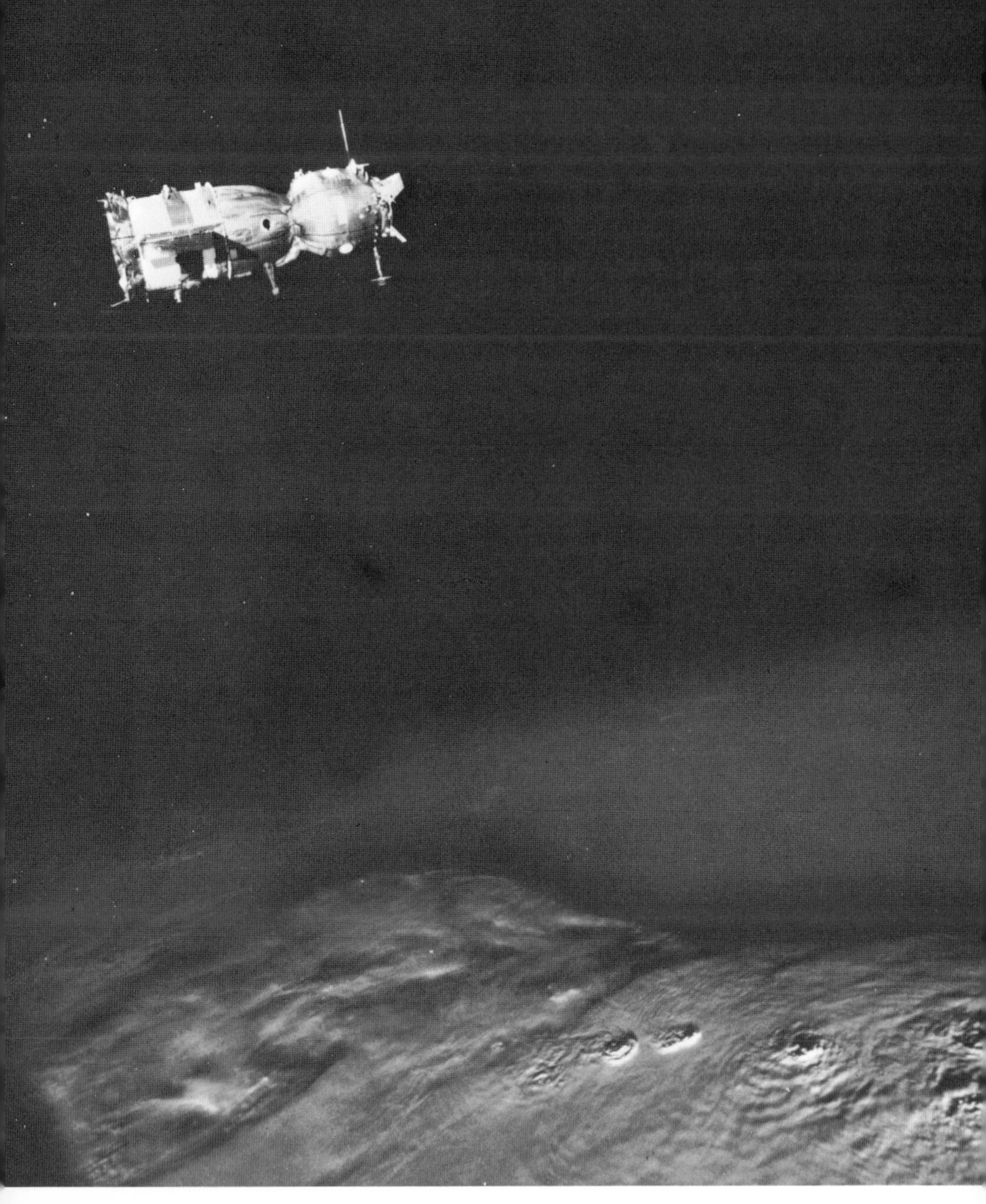

Fig. 10.5. USSR *Soyuz* satellite photographed from the American *Apollo* satellite during rendezvous manoeuvres. A heavily cloud-covered portion of the earth's surface can be seen below, illuminated by the last rays of the setting sun.

could be made from *Soyuz.* Following this experiment the two spacecraft were joined together again briefly before the final undocking and the separate trips back home. The *Soyuz* cosmonauts landed in a dusty field in Kazakstan on 21 July, the first Soviet manned satellite landing to be watched by the outside world on television. *Apollo* tarried a little longer to do more experiments. Its astronauts splashed down in the Pacific Ocean on 24 July. From the American point of view, the mission was marred only by the failure of one of the astronauts to throw a switch that should have deactivated small external rocket thrusters as the re-entry vehicle descended into the atmosphere. Thus, when ports were opened to admit outside air, traces of toxic rocket exhaust fumes came in and caused the astronauts some lung and bronchial irritation from which, fortunately, they now seem to have fully recovered.

Cosmonauts Leonov and Kubasov, and astronauts Stafford, Slayton, and Brand visited each other's spacecraft, ate each other's food, exchanged gold medals, flags, and camaraderie, and held televised press conferences in orbit, designed to convey to the world that 18 years of space competition between the United States and the Soviet Union had finally evolved into co-operation. They also did some experiments which, in addition to the solar coronal photography previously referred to, included ultraviolet and X-ray astronomy, measurement of anomalies in the earth's gravitational field by Doppler velocity measurements between two spacecraft, measurement of atomic oxygen and nitrogen by ultraviolet absorption between *Apollo* and *Soyuz*, observation of the biological effects of high-energy particles on human cells, a study of bacterial transfer between humans in the spacecraft environment, and a number of technological experiments related to weightlessness. Results from most of these experiments had not been published when this was written; however, one result—the discovery of a powerful new stellar source of extreme ultraviolet—was already arousing interest among astronomers.

It was a relatively inexpensive mission (costing the United States a bit less than $240 million and the Soviet Union presumably about the same) because it did not require the development and man-rating of major new spacecraft or launching rockets. Out of it will certainly have come some useful scientific data, some further technological experience, the development of a joint rescue capability by the United States and the Soviet Union for each other's astronauts in distress, if that should ever be needed, and a precedent for active technological co-operation that, even if limited, may be of value for the future.

11 Thinking about the future

If you do not think about the future,
you cannot have one.
John Galsworthy (1928) *Swan song*, Part II, Chapter 2

Only 25 years ago the very idea of creating an artificial satellite to orbit the earth was just a dream in a few imaginative minds. No one could then have predicted most of the accomplishments described in this book. What can now be said of the next few decades? How can we avoid being either too optimistic or too conservative? Perhaps it is best to start with the period only five to ten years ahead.

Short-term prospects

In communications, navigation, meteorology, earth observations, and many areas of science and astronomy, scientists and engineers are eager to press onward, confident that they know how to achieve something even more useful for mankind than has already been achieved, or how to discover new secrets of Nature and the Universe. A few of the goals that are clearly within reach during the next decade are:

- unlimited-capacity, low-cost communication from any point on earth to any other
- the ability to broadcast television with audio commentary in a wide choice of languages or dialects directly into widely dispersed homes, schools, and villages
- two-way video, voice, and data communication between medical clinics in remote villages and a central hospital, so that experienced physicians can diagnose uncommon diseases and direct difficult surgery from a distance
- guidance of ships and aircraft safely over optimum routes, far from land
- better long-range weather forecasting and the beginning of a capability to predict climatic changes
- more comprehensive and timely information about fresh water, minerals, agriculture, forests, and fish, so that these resources can better be managed for the long-range benefit of mankind
- greatly improved understanding of the physics of the sun and of the solar–terrestrial system
- significant progress toward answering some of the ultimate questions of astrophysics, such as the nature of gravity, the origin and evolution of the Universe, the interaction of matter and antimatter, and the energy source for incredibly bright distant objects
- development of useful physical processes which depend on weightlessness
- the acquisition of sufficient medical experience with human beings in a space environment to confidently design manned spaceships for missions of two or three years duration.

Many of these goals will be achieved. However, it is not reasonable to assume that just because a goal is technologically feasible, scientifically interesting, or potentially beneficial in a practical sense, society will automatically assign the necessary resources to accomplish it. Increasingly the word 'priorities' will be heard. The scientific community will not necessarily support a bigger space budget at the expense of ground-based science and astronomy. Even among space enthusiasts, there will be competition between those who want more and bigger satellites and those who want more and bigger spacecraft with which to explore the solar system. Among the general population of voters and taxpayers, lower prices for food, fuel, clothing, transportation, and entertainment, government-subsidized medical care, and low rents or plenty of low-interest mortgage money for housing construction will have the highest priorities. National prestige and other intangible ideological factors may also be important, especially if supported in an emotional way by skilled politicians.

Thus in trying to predict which of many feasible and attractive satellite programmes will actually be supported, we must look first to those that seem to provide a present benefit that is worth more than its cost. Communication satellites are clearly in this class. Earth-observation satellites are also likely to meet the test, but new meteorological satellite developments may be marginal until or unless some dramatic further improvement in forecasting should occur. People tend to remember the few times when the forecast is wrong better than the many times when it is right! As for science and astronomy, the public will probably continue to provide modest support, as in the past, unless the economic situation should become very bad indeed. This support will be given without much understanding of specific scientific objectives, on the basis of a general feeling that scientific achievement is desirable. To a large extent, scientists themselves will have to decide and advise their governments as to priorities between ground-based and space science and between various scientific objectives. Within reasonable limits of objectivity, government leaders will naturally tend to select the more dramatic projects for reasons of prestige and political appeal, but only if the cost is considered to be moderate and bearable.

The shuttle

One obviously desirable alternative is to try to reduce the cost of satellites and of putting them into orbit. Since the beginning, satellites and spacecraft have been launched by large, incredibly expensive machines that have been thrown away after each use. The situation is comparable to discarding a large jet aircraft after just one trip across the ocean. Think of what this would do to the cost of air travel! Engineers have argued, therefore, that satellite launchers should be designed to be fully, or at least partially recoverable, so that their cost can be divided among 25 or 30 missions instead of being chargeable to only one. There are some serious technological problems, but none that seem insoluble. The development cost will be high, but after it will have been paid, the cost of putting a payload into orbit and recovering it again can hopefully be reduced to only a few hundred dollars per kilogram if there will be enough payloads to fully utilize this space transportation capability. After years of study, the United States has decided to take the risk and go ahead with the development of a partially recoverable, manned space transportation vehicle, called simply the *Shuttle.* Estimates indicate that, including the practical-applications satellites, scientific satellites, and military satellites, enough payloads will be available to make this development worthwhile. A significant cost benefit is expected to accrue through

Fig. 11.1. Model of the proposed *Shuttle* with the orbiter, containing a *Spacelab* payload in the foreground, and the large auxilliary tank and solid rocket motors behind.

the reduced costs of the payloads themselves, as they will no longer be subject to the same size and weight constraints, and can be expected to be tested, repaired, and refurbished in orbit, or possibly even recovered and used again.

A model of the *Shuttle* is shown in Fig. 11.1. It consists or an orbiter which looks like a delta-wing aircraft and is in fact about the size of a DC-9 jet transport, plus a very large auxilliary liquid-propellant tank and two solid-propellant rockets. The overall dimension, from the base of the rocket nozzles to the tip of the propellant tank, is about 60 metres. The payload bay, open in this photograph, can accommodate payloads up to about 4·6 metres in diameter and 18·3 metres long. Shown in it are a pressurized laboratory module and an instrument pallet similar to some of those planned for the European *Spacelab* Programme, formerly referred to as *Sortie Lab.* A total payload weight of about 29 500 kilograms can be carried into low-altitude orbit when launched to the eastward from Florida, or about 18 000 kilograms into a low polar orbit. The launch is to be vertical with both solid-propellant rockets and the three orbiter liquid-propellant rocket motors burning. At a speed of about 1370 m/s the two solid-propellant motors will burn out; the empty cases will then be released and descend on parachutes into a nearby ocean area, where they can be recovered and re-used. The liquid-propellant tank will be jettisoned later, outside the atmosphere, in such a way that it will re-enter over a remote ocean area and will not be recoverable. The orbiter, using internally stored propellants, will then manoeuvre to accomplish its mission. If desired, it can remain in orbit for more than a week.

After completion of its mission, the orbiter will use its rocket motors to put it on a re-entry path, enter the atmosphere in a shallow glide, manoeuvre to find its intended landing area, and finally make an unpowered landing like an aircraft with engines off. Thermal protection during re-entry will be provided by a re-useable refractory material covering large areas of the structure, and by the use of carbon-composite material on the leading edges. The orbiter can, if desired, bring as much as 11 500 kilograms of payload back to earth. Flight control will be primarily autonomous—not from the ground. The normal crew of four will include a mission commander, a pilot, a mission specialist, and a payload specialist. Six passengers can be accommodated in the lower deck during launch and recovery, and on some missions two to four scientists may be carried as part of the payload in a pressurized *Spacelab* module in the cargo bay. Fig. 11.2 is an artist's impression of the *Shuttle* orbiter, fitted with such a *Spacelab* module (indicated by the arrow) and two astronomical experiments. Although no details are available, it is understood that the Soviet Union is developing a space transportation vehicle that will be at least functionally similar to the *Shuttle.*

Assuming that these developments will be continued and will be successful, most satellite missions during the 1980s will be launched by the *Shuttle* or its Soviet equivalent. Those that require a higher orbit—geosynchronous satellites, for example—will need additional propulsion, which can be built into the satellites themselves or provided by a standardized orbit transfer stage, sometimes referred to as a *Tug.* The original plan was to make this transfer vehicle also re-useable and it was hoped that it might be developed in Europe. For various reasons this was not possible. An interim expendable transfer stage derived from the upper stage of an existing launch vehicle will therefore be used during the early years of *Shuttle* operations. Availability of low-cost-per-kilogram man-rated space transportation will encourage a trend toward the laboratory-in-space concept exemplified by the United States *Skylab* and

Fig. 11.2. *Shuttle* orbiter with *Spacelab* pressurized laboratory module, shown by arrow, and two astronomy experiments (artist's impression).

the Soviet *Salyut* vehicles. It will also permit the launching of large expensive satellites that preferably would not be occupied continuously by human operators but only visited occasionally for adjustment, replenishment of expendable materials, recovery of film, and the like.

Large space telescope

One such satellite, already being planned for *Shuttle* launch during the early 1980s, is the large space telescope (LST), which was described briefly on p. 144. Fig. 11.3 is an artist's impression of what it will look like. It is expected to be about 12·4 metres long and to have a mass of about 9000 kilograms. The primary mirror will have a diameter of 3 metres and will be located in the after portion of the cylinder, near the ring that supports the solar-cell arrays. The secondary mirror is shown through the cut-away section of the cylinder. In order to achieve the full potential of this instrument, extremely close dimensional tolerances (fractions of a micrometre) must be maintained on the optical parts. There will be no gravitational stress, of course, once the telescope is in orbit. Differential thermal expansion may be the biggest design problem. It is estimated, for example, that a three-degree temperature gradient around the ring holding the primary mirror is more than can be tolerated. Pointing must be steady within a few thousandths of second of arc over the duration of an observation, which may be several hours. Stray light from the sun must be

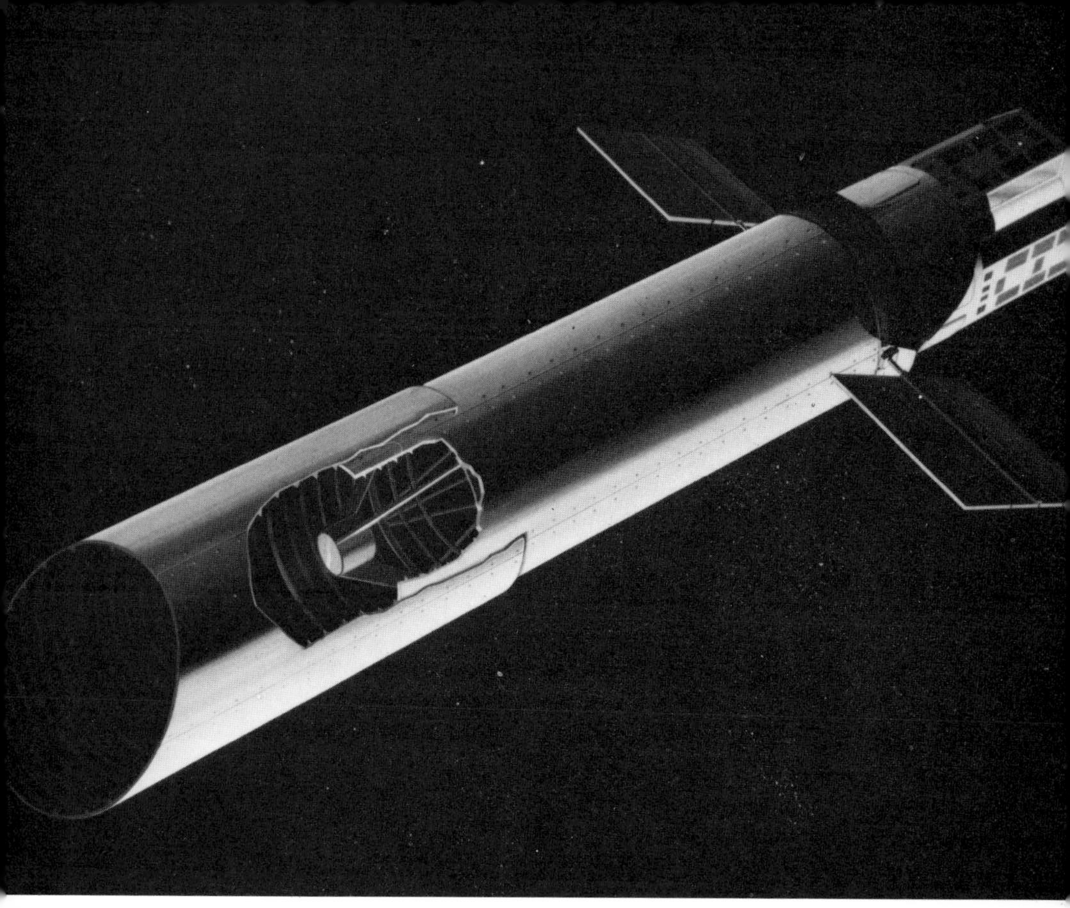

Fig. 11.3. Large space telescope (artist's impression).

reduced to the greatest possible extent. These are extremely difficult engineering problems.

The LST will not be designed to accommodate a full-time live astronomer because his moving around might disturb the pointing accuracy, provisions for his comfort would make the thermal problems more difficult to solve, and occasional venting of cabin atmosphere, water, or waste products might scatter sunlight into the instrument and contaminate optical surfaces. Instead it will be designed to facilitate servicing and adjustment by *Shuttle* crew members who will visit from time to time. Where possible all instruments, electronics, and cameras will be modularized so that an astronaut in a spacesuit will be able to install and remove them from outside the telescope.

The idea of orbiting a large space telescope has been considered previously by both the United States and Europe, but not very seriously because of its high cost. Now, with the *Shuttle*, it seems feasible. Other large astronomical projects, such as a high-resolution kilometre-wave radio-astronomy antenna and a cosmic-ray experiment using a large superconducting magnet, are being thought about.

The more distant future

It would be tempting to dream about giant space projects like a hotel in orbit, where jaded tourists could go to see and experience something really 'different'. However, it is doubtful that anything like that will be done while some people on earth remain poorly housed and fed. If we try to look rationally at the next century, certain high-priority needs appear, and although satellite technology cannot be itself satisfy any of these needs, it may be able to help.

The most important requirement is to limit the population of the world. Already there are ominous signs that the earth cannot support a much larger population than it now has. Many different means for controlling human reproduction are available, but it is necessary in some parts of the world to persuade people to use them. As I suggested earlier, television broadcast satellites could provide a mass-communication capability that might help to accomplish this task.

In the historic past, wars played their part in limiting the numbers of people living on this planet. But even if there were no moral repugnance to this kind of population control, it would no longer be effective. Modern warfare is fought by machines, and it consumes so much of the earth's irreplaceable fuel and mineral resources that it tends to reduce the carrying capacity of the earth more than it reduces the population. Clearly this waste must be ended. Satellite technology may be able to help end wars by providing an international observation capability to police compliance with disarmament agreements.

Our supply of fossil fuels, on which present civilization is based, is like a bank account from which withdrawals are being made at an increasing rate but into which nothing is being deposited. The balance will surely fall to zero within a century or two, and people will be forced to use some other form of energy. Nuclear fission also uses fuel which is available only in limited amounts, but nuclear fusion, if we can learn how to harness it for the generation of electric power, would use fuels that are virtually inexhaustible. That is one reason why scientists are so interested in studying the sun, for it is 'powered' by fusion reactions on a vast scale. Another answer may be to use the sun's radiant energy directly to meet current human needs. One way to do this might be to erect very large solar collectors or arrays of solar cells in orbit and to transmit the energy thus generated down to earth by microwave beams.

In order to provide an ample and reliable supply of food it will be necessary to plan and manage agriculture, animal husbandry, and fisheries more effectively than has been done in the past. It is now clear that the use of satellites to gather information can be of great value for this purpose. It will probably be possible during the next few decades also to provide meaningful climatic forecasts, region by region, as much as a year ahead. At some time during the next century, confidence in climatological understanding may grow to the point where large engineering projects can be undertaken to make arable land out of what is now arid desert or frozen tundra, by modifying the climate. With adequate weather and climate control, disasterous floods and droughts could become spectres of the past.

Finally, it will be necessary to make sure that the human environment does not become irreparably contaminated by the wastes of human activity before corrective action can be taken. To this end, continual monitoring of the atmosphere and oceans by satellites will be indispensible.

The future, of course, will be what people make it. If there exist the intelligence and the will to use wisely the tools that now lie at hand, not the least of which is the satellite technology described in this book, men and women can build a future that will be better than the past.

Appendix. Motion of a small point mass subject to an inverse square force of attraction

For the more mathematically inclined reader, the elementary dynamic equations for a satellite will be solved here in such a way as to show that the orbits must be elliptical and to derive some other interesting characteristics of satellite motion.

Let the satellite be represented by a point mass m and the planetary body in the gravitational field of which the satellite orbits be represented by another but obviously much much greater point mass M. This is admittedly an idealization; such a representation would be exact only if the mass of the planet were distributed in such a way as to have perfect spherical symmetry. In order to further simplify the problem, let it be assumed that M is so much greater than m, that the motion of the planet is unaffected by gravitational attraction of the satellite, and also that no other gravitational influences, such as the sun, the moon, or the other planets, have a significant effect on either the satellite or the planet in question.

The position and velocity of m with respect to a polar coordinate system affixed to M are shown in Fig. A.1. The equations of motion in this coordinate system are then as follows:

$$m\left\{\frac{\mathrm{d}^2 r}{\mathrm{d}t^2} - r\left(\frac{\mathrm{d}\theta}{\mathrm{d}t}\right)^2\right\} = -G\frac{mM}{r^2}, \tag{1}$$

$$m\left\{2\frac{\mathrm{d}r}{\mathrm{d}t}\frac{\mathrm{d}\theta}{\mathrm{d}t} + r\frac{\mathrm{d}^2\theta}{\mathrm{d}t^2}\right\} = 0, \tag{2}$$

where $-GmM/r^2$ is the gravitational force of attraction. Note that if (2) is multiplied by r it becomes

$$m\left\{2r\frac{\mathrm{d}r}{\mathrm{d}t}\frac{\mathrm{d}\theta}{\mathrm{d}t} + r^2\frac{\mathrm{d}^2\theta}{\mathrm{d}t^2}\right\} = m\frac{\mathrm{d}}{\mathrm{d}t}\left(r^2\frac{\mathrm{d}\theta}{\mathrm{d}t}\right) = 0,$$

of which the integral is

$$mr^2\frac{\mathrm{d}\theta}{\mathrm{d}t} = h, \tag{3}$$

in which h is a constant of integration. Now, mr^2 is, by definition, the moment of inertia of the satellite about point M, and $\mathrm{d}\theta/\mathrm{d}t$ is the angular velocity, so (3) states that the angular momentum of a satellite about the planetary centre of mass has a constant value, with which the constant of integration can be identified.

In order to find the equation for the orbit, it is necessary to eliminate the variable t from (1) and (3). As a temporary convenience, introduce a new variable $u = 1/r$. Then, from (3),

$$\frac{\mathrm{d}\theta}{\mathrm{d}t} = \frac{h}{m}u^2$$

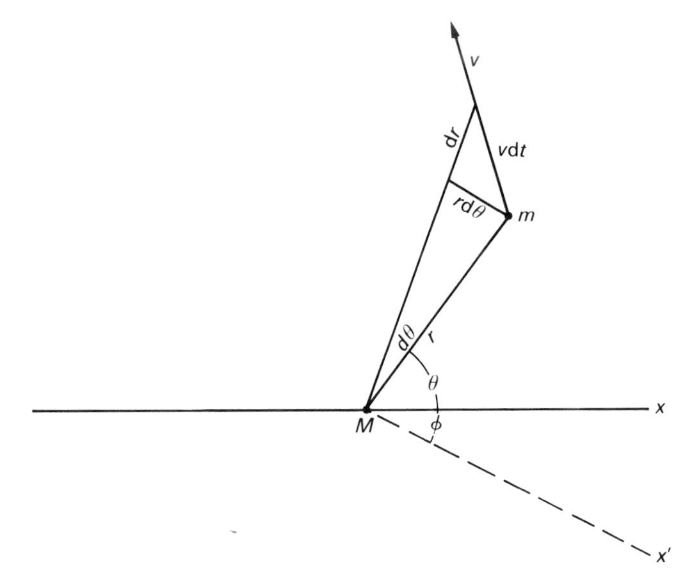

Fig. A.-1. Motion of point mass in time dt.

and

$$\frac{dr}{dt} = \frac{dr}{d\theta}\frac{d\theta}{dt} = \frac{h}{m}\left(\frac{1}{r^2}\frac{dr}{d\theta}\right) = -\frac{h}{m}\frac{du}{d\theta},$$

$$\frac{d^2r}{dt^2} = -\frac{h}{m}\frac{d^2u}{d\theta^2}\frac{d\theta}{dt} = -\left(\frac{h}{m}\right)^2 u^2 \frac{d^2u}{d\theta^2}.$$

Substituting these expressions in (1), it becomes

$$\frac{d^2u}{d\theta^2} + u = \frac{GM}{(h/m)^2}. \tag{4}$$

The complete solution of this differential equation is

$$u = \frac{GM}{(h/m)^2} + A\cos(\theta + \phi), \tag{5}$$

where A and ϕ are constants of integration. Restoring u to its original form of $1/r$, and measuring θ from the x'-axis instead of the x-axis, (5) can be restated as

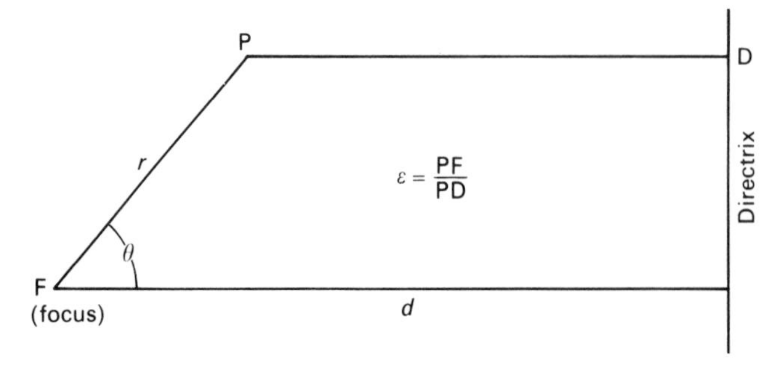

Fig. A.-2. Definition of conic section.

$$\frac{1}{r} = \frac{GM}{(h/m)^2} + A\cos\theta . \tag{6}$$

By definition, a conic section is a curve traced by a point P (Fig. A.2) which moves so that the ratio of its distance from a fixed point F to its distance from a straight line D is constant. The point F is referred to as a *focus*, the line as a *directrix*, and the ratio PF to PD, usually designated by ϵ, as the *eccentricity* of the conic. If d is the distance of the focus from the directrix,

$$\epsilon = \frac{r}{d - r\cos\theta},$$

$$\frac{1}{r} = \frac{1}{\epsilon d} + \frac{1}{d}\cos\theta . \tag{7}$$

Comparing (7) with (6), it is apparent that the motion of m must follow a conic section of eccentricity

$$\epsilon = A\,\frac{(h/m)^2}{GM} \tag{8}$$

and the constant A can be identified as $1/d$. If ϵ is less than one, the conic section must be closed, that is, r must have a finite positive value for any value of θ. The only closed conic section is an ellipse and, of course, a circle, which may be considered to be a special case of an ellipse with $\epsilon = 0$. By definition a satellite must follow a closed orbit—it is so named because it continues to move around its primary body for a long time instead of flying off into space. Therefore, satellite orbits must be ellipses or circles and

$$0 \leqslant \epsilon < 1{\cdot}0 .$$

In order to obtain the energy of the satellite, multiply (1) by $\mathrm{d}r/\mathrm{d}t$, (2) by $r(\mathrm{d}\theta/\mathrm{d}t)$ and add,

$$m\left\{\frac{\mathrm{d}r}{\mathrm{d}t}\frac{\mathrm{d}^2r}{\mathrm{d}t^2} - r\frac{\mathrm{d}r}{\mathrm{d}t}\left(\frac{\mathrm{d}\theta}{\mathrm{d}t}\right)^2 + 2r\frac{\mathrm{d}r}{\mathrm{d}t}\left(\frac{\mathrm{d}\theta}{\mathrm{d}t}\right)^2 + r^2\frac{\mathrm{d}\theta}{\mathrm{d}t}\frac{\mathrm{d}^2\theta}{\mathrm{d}t^2}\right\} = -\frac{GMm}{r^2}\frac{\mathrm{d}r}{\mathrm{d}t},$$

$$m\left\{\frac{\mathrm{d}r}{\mathrm{d}t}\frac{\mathrm{d}^2r}{\mathrm{d}t^2} + r\frac{\mathrm{d}r}{\mathrm{d}t}\left(\frac{\mathrm{d}\theta}{\mathrm{d}t}\right)^2 + r^2\frac{\mathrm{d}\theta}{\mathrm{d}t}\frac{\mathrm{d}^2\theta}{\mathrm{d}t^2} + \frac{GM}{r^2}\frac{\mathrm{d}r}{\mathrm{d}t}\right\} = 0 . \tag{9}$$

The integral of (9) is

$$\frac{m}{2}\left(\frac{\mathrm{d}r}{\mathrm{d}t}\right)^2 + \frac{m}{2}r^2\left(\frac{\mathrm{d}\theta}{\mathrm{d}t}\right)^2 - \frac{GMm}{r} = U, \tag{10}$$

where U is a constant of integration.

Simplifying (10),

$$\frac{m}{2}\left\{\left(\frac{\mathrm{d}r}{\mathrm{d}t}\right)^2 + r^2\left(\frac{\mathrm{d}\theta}{\mathrm{d}t}\right)^2\right\} - \frac{GMm}{r} = U \tag{11}$$

or

$$\frac{1}{2}mv^2 - \frac{GMm}{r} = U, \tag{12}$$

in which $\frac{1}{2}mv^2$ can be identified as the instantaneous kinetic energy of the satellite, GMm/r is instantaneous potential energy, and U the total energy, which is seen to be constant.

Now when m is at its closest approach to M, by definition $\mathrm{d}r/\mathrm{d}t = 0$ and $\theta = 0$. Let the value of r for this situation, known as perigee in the case of an earth satellite, be R_p. Then from (3)

$$R_\mathrm{p}\frac{\mathrm{d}\theta}{\mathrm{d}t} = \frac{(h/m)}{R_\mathrm{p}},$$

and from (6)

$$\frac{1}{R_\mathrm{p}} = \frac{GM}{(h/m)^2} + A\,.$$

$$R_\mathrm{p}\frac{\mathrm{d}\theta}{\mathrm{d}t} = \frac{GM}{(h/m)} + A\,\frac{h}{m}$$

in this situation, where $\mathrm{d}r/\mathrm{d}t = 0$, (11) becomes

$$\left(R_\mathrm{p}\frac{\mathrm{d}\theta}{\mathrm{d}t}\right)^2 - 2\frac{(GM)^2}{(h/m)^2} - 2GMA = \frac{2U}{m},$$

$$\left\{\frac{GM}{(h/m)} + A\left(\frac{h}{m}\right)\right\}^2 - 2\frac{(GM)^2}{(h/m)^2} - 2GMA = \frac{2U}{m},$$

$$\frac{(GM)^2}{(h/m)^2} + 2GMA + A^2\left(\frac{h}{m}\right)^2 - 2\frac{(GM)^2}{(h/m)^2} - 2GMA = \frac{2U}{m},$$

$$-\frac{(GM)^2}{(h/m)^2} + A^2\left(\frac{h}{m}\right)^2 = \frac{2U}{m},$$

which can be written in the form

$$\frac{2U}{m} = -\frac{(GM)^2}{(h/m)^2}\left[1 - \left\{\frac{(h/m)^2 A}{GM}\right\}^2\right]. \tag{13}$$

Substituting the value of ϵ, from (8)

$$\frac{2U}{m} = -\frac{(GM)^2}{(h/m)^2}\left(1 - \epsilon^2\right) \tag{14}$$

or

$$\epsilon = \sqrt{\left\{1 + \frac{2U}{m}\frac{(h/m)^2}{(GM)^2}\right\}}\,. \tag{15}$$

This relationship shows that, in order for ϵ to be less than 1·0, as is required for a closed or elliptical orbit, U must be negative, that is, less than the potential energy of the satellite at an infinite distance from the planet.

An ellipse is usually characterized by its largest diameter, or major axis, as well as by its eccentricity. In (15) the eccentricity of the orbit is related to the energy and angular momentum per unit mass, which are known to be constant for any particular orbit. It is now desired to derive a similar relationship between the major axis, which will be called $2a$, and these same two physical quantities. By definition

$$2a = R_a + R_p = d\left(\frac{\epsilon}{1-\epsilon} + \frac{\epsilon}{1+\epsilon}\right), \tag{16}$$

where R_a and R_p are the maximum and minimum distances (apogee and perigee) from the planetary centre, respectively, and d and ϵ are as defined as in Fig. (A.2). From (16)

$$a = \frac{\epsilon d}{1-\epsilon^2},$$

$$\frac{1}{a} = \frac{1}{\epsilon d} - \frac{\epsilon}{d},$$

and, recalling that $1/d = A$,

$$\frac{1}{a} = \frac{A}{\epsilon} - A\epsilon.$$

But from (8),

$$\frac{A}{\epsilon} = \frac{GM}{(h/m)^2}. \tag{17}$$

Thus

$$\frac{1}{a} = \frac{GM}{(h/m)^2} - \frac{(h/m)^2}{GM}A^2$$

$$= \frac{GM}{(h/m)^2}\left[1 - \left\{\frac{(h/m)^2 A}{GM}\right\}^2\right]. \tag{18}$$

Comparing this with the energy equation in the form of (13), it is apparent that

$$\frac{U}{m} = -\frac{GM}{2a} \tag{19}$$

or

$$a = -\frac{GM}{2(U/m)}. \tag{20}$$

Thus the major axis of the ellipse depends only on the energy per unit mass of the satellite orbit and not on the angular momentum. Also, it is apparent from (19) that the energy of a satellite orbit depends only on the major axis of that orbit, and not at all on the eccentricity or on the minor axis. This is a rather surprising result.

One final characteristic of satellite motion that will be considered here is the period of the orbit, or the time required for the radius vector from M to m to rotate through an angle of 2π radians. Going back to (3), which can be written

$$\frac{1}{2}r\left(r\,\frac{d\theta}{dt}\right) = \frac{h}{2m}, \tag{21}$$

it is evident that the left-hand side is the rate at which area is described ('swept out') by the radius vector and the right-hand side is a constant equal to one-half the angular momentum per unit mass of the satellite.

The area of an ellipse is well known to be $\pi a^2\sqrt{(1-\epsilon^2)}$. Thus, if P is the period of the orbit,

$$P\frac{1}{2}\left(\frac{h}{m}\right) = \pi a^2\sqrt{(1-\epsilon^2)}\,,$$

$$P = \frac{2\pi a^2}{(h/m)}\sqrt{(1-\epsilon^2)}\,. \tag{22}$$

From (18) and (17),

$$\frac{1}{a} = \frac{GM}{(h/m)^2}(1-\epsilon^2)\,,$$

$$\sqrt{(1-\epsilon^2)} = \frac{(h/m)}{\sqrt{(GMa)}}\,.$$

Thus

$$P = \frac{2\pi a^{3/2}}{(GM)^{1/2}}\,. \tag{23}$$

This states that the period of a satellite orbit depends only on the major axis of the orbit and the planetary mass. Therefore an astronomer can determine the mass of any planet which has a satellite (natural or artificial) by measuring only the major axis and period of that satellite's orbit.

Early in the seventeenth century, Johannes Kepler, using the observational data of Tycho Brahe and others, formulated three general laws of planetary motion, which includes the motion of satellites about a planet as well as that of planets about the sun. These three laws, briefly stated were as follow: (1) The orbit is always an ellipse with the sun (or the centre of the planet, in the case of a satellite) as one of the foci; (2) the radius vector from the sun to the planet (or from the centre of a planet to its satellite) always sweeps over equal areas in equal times; and (3) the ratio of the squares of the periods of any two orbits is equal to the ratio of the cubes of the average distances of the planets from the sun (or of the satellites from the centre of the planet). Later, in his famous *Principia.* Isaac Newton showed that these laws could be derived from his more general laws of motion and gravitational attraction, as has been done in this Appendix. The first law is contained in eqn (6), the second in (21), and the third in (23).

Index